I0063578

Update in Pediatric Neuro-Oncology

Update in Pediatric Neuro-Oncology

Special Issue Editors

Soumen Khatua
Natasha Pillay Smiley

MDPI • Basel • Beijing • Wuhan • Barcelona • Belgrade

MDPI

Special Issue Editors
Soumen Khatua
The University of Texas MD Anderson Cancer Center
USA

Natasha Pillay Smiley
Ann and Robert H. Lurie Children's Hospital
USA

Editorial Office
MDPI
St. Alban-Anlage 66
4052 Basel, Switzerland

This is a reprint of articles from the Special Issue published online in the open access journal *Bioengineering* (ISSN 2306-5354) in 2018 (available at: https://www.mdpi.com/journal/bioengineering/special_issues/pediatr_neuro_oncol)

For citation purposes, cite each article independently as indicated on the article page online and as indicated below:

LastName, A.A.; LastName, B.B.; LastName, C.C. Article Title. *Journal Name* **Year**, *Article Number, Page Range.*

ISBN 978-3-03897-539-7 (Pbk)
ISBN 978-3-03897-540-3 (PDF)

Cover image courtesy of Natasha Pillay Smiley.

ⓒ 2019 by the authors. Articles in this book are Open Access and distributed under the Creative Commons Attribution (CC BY) license, which allows users to download, copy and build upon published articles, as long as the author and publisher are properly credited, which ensures maximum dissemination and a wider impact of our publications.

The book as a whole is distributed by MDPI under the terms and conditions of the Creative Commons license CC BY-NC-ND.

Contents

About the Special Issue Editors vii

Preface to "Update in Pediatric Neuro-Oncology" ix

Natasha Pillay Smiley and Soumen Khatua
Introduction to the Special Issue on Pediatric Neuro-Oncology
Reprinted from: *Bioengineering* **2018**, *5*, 109, doi:10.3390/bioengineering5040109 1

Anders W. Bailey, Amreena Suri, Pauline M. Chou, Tatiana Pundy, Samantha Gadd,
Stacey L. Raimondi, Tadanori Tomita and Simone Treiger Sredni
Polo-Like Kinase 4 (PLK4) Is Overexpressed in Central Nervous System Neuroblastoma
(CNS-NB)
Reprinted from: *Bioengineering* **2018**, *5*, 96, doi:10.3390/bioengineering5040096 5

Amer M. Najjar, Jason M. Johnson and Dawid Schellingerhout
The Emerging Role of Amino Acid PET in Neuro-Oncology
Reprinted from: *Bioengineering* **2018**, *5*, 104, doi:10.3390/bioengineering5040104 18

Ethan B. Ludmir, David R. Grosshans and Kristina D. Woodhouse
Radiotherapy Advances in Pediatric Neuro-Oncology
Reprinted from: *Bioengineering* **2018**, *5*, 97, doi:10.3390/bioengineering5040097 33

Cavan P. Bailey, Mary Figueroa, Sana Mohiuddin, Wafik Zaky and Joya Chandra
Cutting Edge Therapeutic Insights Derived from Molecular Biology of Pediatric High-Grade
Glioma and Diffuse Intrinsic Pontine Glioma (DIPG)
Reprinted from: *Bioengineering* **2018**, *5*, 88, doi:10.3390/bioengineering5040088 49

Peter H. Baenziger and Karen Moody
Palliative Care for Children with Central Nervous System Malignancies
Reprinted from: *Bioengineering* **2018**, *5*, 85, doi:10.3390/bioengineering5040085 65

Tara H.W. Dobson and Vidya Gopalakrishnan
Preclinical Models of Pediatric Brain Tumors—Forging Ahead
Reprinted from: *Bioengineering* **2018**, *5*, 81, doi:10.3390/bioengineering5040081 83

David E. Kram, Jacob J. Henderson, Muhammad Baig, Diya Chakraborty,
Morgan A. Gardner, Subhasree Biswas and Soumen Khatua
Embryonal Tumors of the Central Nervous System in Children: The Era of
Targeted Therapeutics
Reprinted from: *Bioengineering* **2018**, *5*, 78, doi:10.3390/bioengineering5040078 96

Peter L. Stavinoha, Martha A. Askins, Stephanie K. Powell, Natasha Pillay Smiley and
Rhonda S. Robert
Neurocognitive and Psychosocial Outcomes in Pediatric Brain Tumor Survivors
Reprinted from: *Bioengineering* **2018**, *5*, 73, doi:10.3390/bioengineering5030073 112

About the Special Issue Editors

Soumen Khatua is an Associate Professor and a pediatric Neuro-Oncologist at M.D. Anderson Cancer Center. He completed a pediatric Hematology-Oncology fellowship at the Children's National Medical Center, Washington DC and a Neuro-Oncology fellowship at the Children's Hospital Los Angeles. His research efforts are directed towards developing clinical trials using targeted therapy in pediatric brain tumors. Dr. Khatua's areas of interest and focus are high-grade glioma, diffuse pontine glioma and intracranial germ cell tumors.

Natasha Pillay Smiley is a pediatric Neuro-Oncologist at Ann & Robert H. Lurie in Chicago, Illinois. In addition to taking care of newly diagnosed children with brain and spinal cord tumors, she serves as the director of the Pediatric Brain Tumor Survivorship Program and is a member of the Cancer Predisposition clinic and the NF-1 Neuro-Oncology Clinic. She also currently serves as the Assistant Hematology/Oncology/SCT Fellowship Director and the Neuro-Oncology Fellowship Director.

Preface to "Update in Pediatric Neuro-Oncology"

Pediatric Neuro-Oncology is a highly specialized field encompassing molecular biology, clinical acumen, evidence-based medicine, cancer genetics, and neuropsychological care for the diagnosis and treatment of children with central nervous system (CNS) tumors. In this Special Edition of Bioengineering, we hope to demonstrate the wide breath of science and medicine that occurs in the field of pediatric neuro-oncology. Faced with substantial mortality in children with aggressive tumors as well as significant morbidity of survivors, we are always challenged to learn more about these disease entities and improve the outcomes of these children. Topics that are discussed further in this edition are: Molecular biology in pediatric gliomas, the clinical relevance of preclinical models, updates on radiation therapy for pediatric CNS tumors, molecular neuro-imaging, embryonal tumors and targeted therapeutics, and neurocognitive and psychosocial outcomes and palliative care in children with central nervous system malignancies.

Soumen Khatua, Natasha Pillay Smiley
Special Issue Editors

bioengineering

MDPI

Editorial

Introduction to the Special Issue on Pediatric Neuro-Oncology

Natasha Pillay Smiley [1],* and Soumen Khatua [2]

[1] Department of Hematology/Oncology/SCT. Ann & Robert H. Lurie Children's Hospital, Chicago, IL 60611, USA;
[2] Department of Pediatrics, The University of Texas MD Anderson Cancer Center, Houston, TX 77030, USA; skhatua@mdanderson.org
* Correspondence: npillaysmiley@luriechildrens.org

Received: 25 November 2018; Accepted: 6 December 2018; Published: 11 December 2018

Keywords: brain tumor; pediatrics; advancements; molecular biology

Pediatric Neuro-Oncology is a highly specialized field encompassing molecular biology, clinical acumen, evidence based medicine, cancer genetics and neuropsychological care for the diagnosis and treatment of children with central nervous system (CNS) tumors. In data acquired by the National Institute of Health's (NIH) Surveillance, Epidemiology and End Results (SEER) Program, there were 3.1 new cases of childhood brain tumors per 100,000 people from 2011–2015. This represents the second most common pediatric cancer diagnosis (17.2% overall, second only to leukemia) as well as the leading cause of mortality [1]. This special edition of the *Bioengineering* Journal highlights major advancements in pediatric neuro-oncology as well as our current challenges.

The SEER data describes the overall survival of pediatric CNS tumors to be approximately 70%, an improvement from less than 60% from the 1970s [1]. This statistic can be misrepresentative as pediatric CNS tumors are heterogenous with diverse survival outcomes, ranging from the mostly indolent nature of low grade gliomas with a 20 year overall survival (OS) of 80% [2] to the highly aggressive diffuse intrinsic pontine glioma (DIPG) with a 2 year OS of less than 10%. [3,4].

Although survival has not dramatically increased for DIPG and high grade glioma over the last few decades, there has been an increase in the survivability of other tumors such as the medulloblastoma (MB) and atypical teratoid rhabdoid tumor (ATRT). This has resulted from the surge of genomic and epigenomic data of pediatric brain tumors- forging an era of biologically targeted therapy with improved survival outcome of CNS tumors. This is best illustrated by the story of medulloblastoma subtyping and the somewhat recent discovery of ATRT. We now have established molecular subgroups with defined demographics, oncogenic drivers and risk stratification based treatment strategies. Retrospective analysis of medulloblastoma patients have determined that children with WNT subgrouping have a significantly higher overall survival, and thus, clinical trials are now focused on changing upfront treatment of these children to mitigate the profound late effects medulloblastoma patients face [5].

Atypical Teratoid Rhabdoid tumor (ATRT) was only described in the 1980s, previously thought to be a type of medulloblastoma or supratentorial PNET [6,7]. Advances in histologic characterization and FISH for chromosome 22 helped to classify this as a separate entity. ATRT is a disease of primarily infants, and was nearly always fatal, with the 3 years survival of children treated with the Pediatric Oncology Group (POG) infant studies of less than 10% [7]. However, treatment with high dose chemotherapy and/or autologous transplantation and radiation has now led to improved survival in this population of young children. The two years overall survival for the DFCI regimen is 70 +/− 10%, using intrathecal chemotherapy, focal or craniospinal radiation and dose intensive chemotherapy [7]. The Vienna regimen, which had a smaller cohort of patients, is also a dose intensive regimen and has an excellent 5 years overall survival of 100% using methotrexate, intrathecal chemotherapy, anthracyclines, focal radiation and autologous transplantation [8]. As expected, there are long term effects from these treatments and children have been found to have neurocognitive sequelae even in the absence of radiation [9]. Molecular subtyping using methylation profiles has now delineated three subtypes of ATRT, with the hope that risk stratification can help further improve survival while decreasing toxicity and long term effects [10].

Arguably, as illustrated above, the most critical advancement in our field is the attainment of an accurate diagnosis, which has implications not only for individual patient care but also for basic science and clinical trial research [11]. The World Health Organization (WHO) Classification of CNS Tumors represents a consensus opinion from world experts and allows pathologists and neuro-oncologists across the world the opportunity to have guidelines to define CNS tumors [12]. A major change occurred in the most recent edition of the guidelines set forth in 2016. Distinct molecular characteristics were integrated into the classification of CNS tumors, allowing for an "integrated diagnosis" that is "layered" with both histologic features and molecular biology [12]. Histologic analysis depends on defining tumors by cell of origin and level of differentiation. This is accomplished by examining "hematoxylin and eosin-stained (H & E) sections, immunohistochemical expression of lineage associated proteins and ultrastructural characterization" [11]. This well-established method is now augmented by molecular analysis of the genotype of these tumors. This change brings scientific advancement into direct patient care and is an example of the innovative nature of this field. Less formally, but perhaps no less important, parameters such as neuro-imaging and clinical course is also taken into account to complete the integrated diagnosis.

Translational research bridges the gap between basic science and cancer treatments. Major advancements in next generation sequencing technology has led to greater understanding of cancer genomes and thus led to potential cures for patients [13]. This is beautifully illustrated in the landscape of pediatric low grade glioma, the most common central nervous system tumor in children [13]. The "integrated diagnosis" now routinely includes histologic grading as well as whether the tumor has particular aberrations in the MAP kinase (MAPK) pathway [13–16]. While multiple genetic changes have been seen in pediatric low grade glioma, the most common involve the MAPK pathway, specifically, either an activating point mutation of BRAFV600E or activating of BRAF through a tandem duplication. This results in the KIAA 1549-BRAF fusion protein [14–16]. Molecular analysis has been correlated with histologic characterization, and 70–90% of pilocytic astrocytomas have been found to have a BRAF-KIAA1549 fusion [15]. In addition, BRAF v600 E has been found to be aberrant in other low grade gliomas such as pilomyxoid astrocytomas [14]. Drugs have been developed to selectively inhibit these targets, and early phase clinical trials have been undertaken to understand their tolerability and efficacy in children. The traditional methods of treating low grade glioma are systemic chemotherapy and, more remotely, radiation. These modalities can cause significant late effects in patients, and maximizing efficacy while minimizing long term effects are important in a population with an expected long term survival [2,15]. A Phase I trial through the Pediatric Brain Tumor Consortium (PBTC) used selumetinib (AZD6244, AstraZeneca), an oral small molecule inhibitor of MEK-1/2, in children with recurrent low grade glioma. A dose was established to perform the phase II trial, in which efficacy will be tested. However, promising antitumor effect was seen in the phase I trial [16].

In this special edition of *Bioengineering*, we hope to demonstrate the wide breath of science and medicine that occurs in the field of pediatric neuro-oncology (Table 1). Faced with substantial mortality in children with aggressive tumors as well as significant morbidity of survivors, we are always challenged to learn more about these disease entities and improve the outcomes of these children. Topics that will be discussed further in this edition are: molecular biology in pediatric gliomas, clinical relevance of preclinical models, update on radiation therapy for pediatric CNS tumors, molecular neuro-imaging, embryonal tumors and targeted therapeutics, neurocognitive and psychosocial outcomes and palliative care in children with central nervous system malignancies.

Table 1. Published papers in Special Issue Update in Pediatric Neuro-Oncology.

Papers	Reference
Preclinical Models of Pediatric Brain Tumors—Forging Ahead	[17]
Cutting Edge Therapeutic Insights Derived from Molecular Biology of Pediatric High-Grade Glioma and Diffuse Intrinsic Pontine Glioma (DIPG)	[18]
Polo-Like Kinase 4 (PLK4) Is Overexpressed in Central Nervous System Neuroblastoma (CNS-NB)	[19]
Embryonal Tumors of the Central Nervous System in Children: The Era of Targeted Therapeutics	[20]
Radiotherapy Advances in Pediatric Neuro-Oncology	[21]
The Emerging Role of Amino Acid PET in Neuro-Oncology	[22]
Palliative Care for Children with Central Nervous System Malignancies	[23]
Neurocognitive and Psychosocial Outcomes in Pediatric Brain Tumor Survivors	[24]

Conflicts of Interest: The authors declare no conflict of interest.

References

1. SEER Database. Available online: https://seer.cancer.gov/statfacts/html/childbrain.html (accessed on 27 November 2018).
2. Bandopadhayay, P.; Bergthold, G.; London, W.B.; Goumnerova, L.C.; Morales La Madrid, A.; Marcus, K.J.; Guo, D.; Ullrich, N.J.; Robison, N.J.; Chi, S.N.; et al. Long-term outcome of 4040 children diagnosed with pediatric low-grade gliomas: An analysis of the Surveillance Epidemiology and End Results (SEER) database. *Pediatr. Blood Cancer.* **2014**, *61*, 1173–1179. [CrossRef]
3. Cohen, K.J.; Pollack, I.F.; Zhou, T.; Buxton, A.; Holmes, E.J.; Burger, P.C.; Brat, D.J.; Rosenblum, M.K.; Hamilton, R.L.; Lavey, R.S.; et al. Temozolomide in the treatment of high-grade gliomas in children: A report from the Children's Oncology Group. *Neuro Oncol.* **2011**, *13*, 317–323. [CrossRef]
4. Jansen, M.H.; Veldhuijzen van Zanten, S.E.; Sanchez Aliaga, E.; Heymans, M.W.; Warmuth-Metz, M.; Hargrave, D.; Van Der Hoeven, E.J.; Gidding, C.E.; de Bont, E.S.; Eshghi, O.S.; et al. Survival prediction model of children with diffuse intrinsic pontine glioma based on clinical and radiological criteria. *Neuro Oncol.* **2015**, *17*, 160–166. [CrossRef]
5. Millard, N.E.; De Braganca, K.C. Medulloblastoma. *J. Child Neurol.* **2016**, *31*, 1341–1353. [CrossRef]
6. Burger, P.C.; Yu, I.T.; Tihan, T.; Friedman, H.S.; Strother, D.R.; Kepner, J.L.; Duffner, P.K.; Kun, L.E.; Perlman, E.J. Atypical teratoid/rhabdoid tumor of the central nervous system: A highly malignant tumor of infancy and childhood frequently mistaken for medulloblastoma: A Pediatric Oncology Group study. *Am. J. Surg. Pathol.* **1998**, *22*, 1083–1092. [CrossRef]
7. Chi, S.N.; Zimmerman, M.A.; Yao, X.; Cohen, K.J.; Burger, P.; Biegel, J.A.; Rorke-Adams, L.B.; Fisher, M.J.; Janss, A.; Mazewski, C.; et al. Intensive Multimodality Treatment for Children with Newly Diagnosed CNS Atypical Teratoid Rhabdoid Tumor. *J. Clin. Oncol.* **2009**, *27*, 385–389. [CrossRef]
8. Slavc, I.; Chocholous, M.; Leiss, U.; Haberler, C.; Peyrl, A.; Azizi, A.A.; Dieckmann, K.; Woehrer, A.; Peters, C.; Widhalm, G.; et al. Atypical teratoid rhabdoid tumor: Improved long-term survival with an intensive multimodal therapy and delayed radiotherapy. The Medical University of Vienna Experience1992–2012. *Cancer Med.* **2014**, *3*, 91–100. [CrossRef]
9. Lafay-Cousin, L.; Fay-McClymont, T.; Johnston, D.; Fryer, C.; Scheinemann, K.; Fleming, A.; Hukin, J.; Janzen, L.; Guger, S.; Strother, D.; et al. Neurocognitive Evaluation of Long Term Survivors of Atypical Teratoid Rhabdoid Tumors (ATRT): The Canadian Registry Experience. *Pediatr. Blood Cancer* **2015**, *62*, 1265–1269. [CrossRef]
10. Jones, D.T.; Kieran, M.W.; Bouffet, E.; Alexandrescu, S.; Bandopadhayay, P.; Bornhorst, M.; Ellison, D.; Fangusaro, J.; Fisher, M.J.; Foreman, N.; et al. Pediatric low-grade gliomas: Next biologically driven steps. *Neuro-Oncol.* **2018**, *20*, 160–173. [CrossRef]
11. Louis, D.N.; Perry, A.; Burger, P.; Ellison, D.W.; Reifenberger, G.; von Deimling, A.; Aldape, K.; Brat, D.; Collins, V.P.; Eberhart, C.; et al. International Society of Neuropathology-Haarlem Consensus Guidelines for Nervous System Tumor Classification and Grading. *Brain Pathol.* **2014**, *24*, 429–435. [CrossRef]

12. Louis, D.N.; Perry, A.; Reifenberger, G.; Von Deimling, A.; Figarella-Branger, D.; Cavenee, W.K.; Ohgaki, H.; Wiestler, O.D.; Kleihues, P.; Ellison, D.W. The 2016 World Health Organization Classification of Tumors of the Central Nervous System: A summary. *Acta Neuropathol.* **2016**, *131*, 803–820. [CrossRef] [PubMed]
13. Meyerson, M.; Gabriel, S.; Getz, G. Advances in understanding cancer genomes through second- generation sequencing. *Nat. Rev. Genet.* **2010**, *11*, 685–696. [CrossRef] [PubMed]
14. Bergthold, G.; Bandopadhayay, P.; Bi, W.L.; Ramkissoon, L.; Stiles, C.; Segal, R.A.; Beroukhim, R.; Ligon, K.L.; Grill, J.; Kieran, M.W. Pediatric low-grade gliomas: How modern biology reshapes theclinical field. *Biochim. Biophys. Acta* **2014**, *1845*, 294–307. [CrossRef] [PubMed]
15. Packer, R.J.; Pfister, S.; Bouffet, E.; Avery, R.; Bandopadhayay, P.; Bornhorst, M.; Bowers, D.C.; Ellison, D.; Fangusaro, J.; Foreman, N.; et al. Pediatric low-grade gliomas: Implications of the biologic era. *Neuro-Oncol.* **2017**, *19*, 750–761. [CrossRef] [PubMed]
16. Banerjee, A.; Jakacki, R.I.; Onar-Thomas, A.; Wu, S.; Nicolaides, T.; Young Poussaint, T.; Fangusaro, J.; Phillips, J.; Perry, A.; Turner, D.; et al. A phase I trial of the MEK inhibitor selumetinib (AZD6244) in pediatric patients with recurrent or refractory low-grade glioma: A Pediatric Brain Tumor Consortium (PBTC) study. *Neuro-Oncol.* **2016**, *19*, 1135–1144. [CrossRef] [PubMed]
17. Dobson, T.H.; Gopalakrishnan, V. Preclinical Models of Pediatric Brain Tumors—Forging Ahead. *Bioengineering* **2018**, *5*, 81. [CrossRef] [PubMed]
18. Bailey, C.P.; Figueroa, M.; Mohiuddin, S.; Zaky, W.; Chandra, J. Cutting Edge Therapeutic Insights Derived from Molecular Biology of Pediatric High-Grade Glioma and Diffuse Intrinsic Pontine Glioma (DIPG). *Bioengineering* **2018**, *5*, 88. [CrossRef] [PubMed]
19. Bailey, A.W.; Suri, A.; Chou, P.M.; Pundy, T.; Gadd, S.; Raimondi, S.L.; Tomita, T.; Sredni, S.T. Polo-Like Kinase 4 (PLK4) Is Overexpressed in Central Nervous System Neuroblastoma (CNS-NB). *Bioengineering* **2018**, *5*, 96. [CrossRef] [PubMed]
20. Kram, D.E.; Henderson, J.J.; Baig, M.; Chakraborty, D.; Gardner, M.A.; Biswas, S.; Khatua, S. Embryonal Tumors of the Central Nervous System in Children: The Era of Targeted Therapeutics. *Bioengineering* **2018**, *5*, 78. [CrossRef] [PubMed]
21. Ludmir, E.B.; Grosshans, D.R.; Woodhouse, K.D. Radiotherapy Advances in Pediatric Neuro-Oncology. *Bioengineering* **2018**, *5*, 97. [CrossRef] [PubMed]
22. Najjar, A.M.; Johnson, J.M.; Schellingerhout, D. The Emerging Role of Amino Acid PET in Neuro-Oncology. *Bioengineering* **2018**, *5*, 104. [CrossRef] [PubMed]
23. Baenziger, P.H.; Moody, K. Palliative Care for Children with Central Nervous System Malignancies. *Bioengineering* **2018**, *5*, 85. [CrossRef] [PubMed]
24. Stavinoha, P.L.; Askins, M.A.; Powell, S.K.; Pillay Smiley, N.; Robert, R.S. Neurocognitive and Psychosocial Outcomes in Pediatric Brain Tumor Survivors. *Bioengineering* **2018**, *5*, 73. [CrossRef] [PubMed]

© 2018 by the authors. Licensee MDPI, Basel, Switzerland. This article is an open access article distributed under the terms and conditions of the Creative Commons Attribution (CC BY) license (http://creativecommons.org/licenses/by/4.0/).

bioengineering

MDPI

Communication

Polo-Like Kinase 4 (PLK4) Is Overexpressed in Central Nervous System Neuroblastoma (CNS-NB)

Anders W. Bailey [1,3,†], **Amreena Suri** [1,3,†], **Pauline M. Chou** [4,5], **Tatiana Pundy** [1], **Samantha Gadd** [4,5], **Stacey L. Raimondi** [6], **Tadanori Tomita** [1,2] **and Simone Treiger Sredni** [1,2,3,*]

1   Division of Pediatric Neurosurgery, Ann and Robert H. Lurie Children's Hospital of Chicago, Chicago, IL 60611, USA; anbailey@luriechildrens.org (A.W.B.); aisuri@luriechildrens.org (A.S.); TPundy@luriechildrens.org (T.P.); TTomita@luriechildrens.org (T.T.)
2   Department of Surgery, Feinberg School of Medicine, Northwestern University, Chicago, IL 60611, USA
3   Cancer Biology and Epigenomics Program, Stanley Manne Children's Research Institute, Chicago, IL 60614, USA
4   Department of Pathology, Ann and Robert H. Lurie Children's Hospital of Chicago, Chicago, IL 60611, USA; PChou@luriechildrens.org (P.M.C.); sgadd@luriechildrens.org (S.G.)
5   Department of Pediatrics, Feinberg School of Medicine, Northwestern University, Chicago, IL 60611, USA
6   Department of Biology, Elmhurst College, Elmhurst, IL 60126, USA; raimondis@elmhurst.edu
*   Correspondence: ssredni@luriechildrens.org or ssredni@northwestern.edu; Tel. +1-773-755-6526
†   Authors with equal contribution.

Received: 28 August 2018; Accepted: 1 November 2018; Published: 4 November 2018

Abstract: Neuroblastoma (NB) is the most common extracranial solid tumor in pediatrics, with rare occurrences of primary and metastatic tumors in the central nervous system (CNS). We previously reported the overexpression of the polo-like kinase 4 (PLK4) in embryonal brain tumors. PLK4 has also been found to be overexpressed in a variety of peripheral adult tumors and recently in peripheral NB. Here, we investigated *PLK4* expression in NBs of the CNS (CNS-NB) and validated our findings by performing a multi-platform transcriptomic meta-analysis using publicly available data. We evaluated the *PLK4* expression by quantitative real-time PCR (qRT-PCR) on the CNS-NB samples and compared the relative expression levels among other embryonal and non-embryonal brain tumors. The relative *PLK4* expression levels of the NB samples were found to be significantly higher than the non-embryonal brain tumors (p-value < 0.0001 in both our samples and in public databases). Here, we expand upon our previous work that detected *PLK4* overexpression in pediatric embryonal tumors to include CNS-NB. As we previously reported, inhibiting PLK4 in embryonal tumors led to decreased tumor cell proliferation, survival, invasion and migration in vitro and tumor growth in vivo, and therefore PLK4 may be a potential new therapeutic approach to CNS-NB.

Keywords: embryonal brain tumor; pediatric; CNS-PNET; low grade glioma; rhabdoid; ATRT; medulloblastoma; kinase inhibitor

1. Introduction

Embryonal tumors of the central nervous system (CNS) are poorly differentiated tumors resembling the developing embryonic nervous system. Embryonal tumors are biologically aggressive and have a tendency to disseminate along cerebrospinal fluid pathways. In the CNS, this group includes medulloblastoma (MB) [1], atypical teratoid/rhabdoid tumor (ATRT) [2], embryonal tumor with multilayer rosettes (ETMR) [3], a spectrum of tumors called "CNS primitive neuroectodermal tumors (PNETs)" and CNS neuroblastoma (CNS-NB) [4].

Neuroblastoma (NB) is the most common extracranial pediatric solid tumor [5]. Current therapies have led to a 90% survival rate, but relapse and metastases have proven to be challenging to treat with survival rates of less than 40% [6].

Previously, we performed a partial functional screening of the kinome on a well-established embryonal tumor cell line (MON—a rhabdoid tumor cell line provided by Dr. Delattre, Institut Curie, Paris France) [7–9] using lentiviral-CRISPR to target 160 individual kinase encoding genes representing the major branches of the human kinome and key isoforms within each branch. With this approach we identified the polo-like kinase 4 (PLK4) as a putative genetic hit. The genetic loss-of-function was validated by next-generation sequencing analysis, genomic cleavage detection (GCD) assay, quantitative real-time PCR (qRT-PCR) and western blot [7]. We established that PLK4 is overexpressed in embryonal brain tumors such as ATRT and MB [10,11]. We also demonstrated that inhibiting PLK4 with the small-molecule inhibitor CFI-400945 (CAS#1338800-06-8) [12–14] resulted in impairment of proliferation, survival, migration and invasion in ATRT and MB cell lines. Further, we demonstrated that PLK4 inhibition induced apoptosis, senescence and polyploidy in these cells. Moreover, we established that polyploidy induced by PLK4 inhibition increased tumor cell susceptibility to DNA-damaging agents while sparing non-tumor cells [7,10].

PLK4 is a cell cycle regulated protein specifically recruited at the centrosome to promote the duplication of centrioles in dividing cells [15–17]. Complete loss of PLK4 is lethal and its overexpression triggers centrosomal amplification, which is associated with genetic instability and consequently, carcinogenesis [18,19]. Active PLK4 protein levels have previously been described to be "mirrored by *PLK4* mRNA levels" meaning that mRNA expression varies proportionally to protein expression [15]. Although PLK4 has been found to be overexpressed in a number of adult peripheral tumors like colorectal [20], breast [21], lung [22], melanoma [23], leukemia [24], and pancreatic cancer [25], we were the first to report PLK4 overexpression in embryonal tumors and in pediatric brain tumors [7,10,11]. Recently, Tian and colleagues reported PLK4 overexpression in peripheral NB tumor samples and primary NB cell lines. They also demonstrated that increased PLK4 expression was correlated with poor clinical outcomes [6]. Here, we hypothesize that, as in other CNS embryonal brain tumors, CNS-NB overexpress PLK4. To test our hypothesis, we examined *PLK4* expression in NB samples of the CNS as compared to other embryonal brain tumors (ATRT and MB) and low grade gliomas (LGG), which are the most common form of primary CNS tumors. For this, we performed quantitative real-time PCR (qRT-PCR) in our patients' tumor samples and an extensive multi-platform transcriptomic meta-analysis using publicly available databases.

2. Materials and Methods

2.1. Quantitative Real-Time PCR (qRT-PCR)

Fresh frozen tumor samples were obtained from the Falk Brain Tumor Bank (Chicago, IL, USA) and the Center for Childhood Cancer, Biopathology Center (Columbus, OH, USA), which is a section of the Cooperative Human Tissue Network of The National Cancer Institute (Bethesda, MD, USA). Written informed parental consents were obtained prior to sample collection. The study was approved by the institutional review board of the Ann and Robert H. Lurie Children's Hospital of Chicago (IRB 2005–12,252; 2005–12,692; 2009–13,778; and 2012–14,887). Samples in the study included 2 CNS-NB (primary $n = 1$ and metastatic $n = 1$), 6 embryonal brain tumors (ATRT $n = 3$ and MB $n = 3$) and 6 non-embryonal brain tumors (low grade gliomas—LGG).

Total RNA was isolated from each frozen tumor sample using TRIzol Reagent (Thermo Fisher, USA). The expression of *PLK4* (Hs00179514_m1) was accessed by TaqMan GE assays (Applied Biosystems, USA). Three housekeeping genes: *GAPDH* (Hs02758991_g1), *HPRT* (Hs99999909_m1) and *HMBS* (Hs00609296_g1) were used as references as previously described [7,10,26–28]. Total RNA (2 µg) was used to make cDNA using the Applied Biosystems High Capacity RNA-to-cDNA kit (Thermo Fisher Scientific, Waltham, MA, USA). Reactions were performed in triplicates with adequate positive and negative controls. The normalized expression levels were calculated by the ΔΔCt method using each housekeeping gene and a pool of all samples as calibrator. The normalized expression levels were also calculated using a normalization factor which was obtained by calculating the geometric mean of relative quantities of all 3 housekeeping genes and dividing the relative quantity of *PLK4* with this normalization factor [7,10,26–28]. Statistical analysis was performed using a One-Way ANOVA using PRISM (GraphPad 7 Software, Inc., La Jolla, CA, USA).

2.2. Gene Expression Meta-Analysis

In order to validate the *PLK4* expression levels observed in our patients, we performed an extensive meta-analysis compiling publicly available gene expression data. Knowing, from our previous studies that *PLK4* is overexpressed in embryonal brain tumors [6,7,11], we selected low grade gliomas (LGG), which are the most common form of primary CNS tumors arising in both children and adults [29,30] to perform this comparison.

To evaluate the *PLK4* expression profile in both tumor and normal human tissues, expression levels of *PLK4* (ENSG00000142731.6) were compared with expression levels of the neuroendocrine marker used for the diagnosis of neuroblastoma chromogranin A (*CHGA*, ENSG00000100604.11) [31] and the glioma markers glial fibrillary acidic protein (*GFAP* ENSG00000131095.10) and myelin basic protein (*MBP*, ENSG00000197971.10) [32,33].

Tumors: Open access transcriptomic data (RNAseqV2, FPKM) from NB samples which were deposited in the TARGET (Therapeutically Applicable Research to Generate Effective Treatments, https://ocg.cancer.gov/programs/target) database and LGG expression data which were deposited in the TCGA (The Cancer Genome Atlas, https://cancergenome.nih.gov/) database, were obtained from the Genomic Data Commons (GDC) (https://portal.gdc.cancer.gov/).

Normal human tissue: Open access transcriptomic data from 51 tissue types represented in the GTEx (Genotype-Tissue Expression, https://gtexportal.org) portal [34] was analyzed. Each gene of interest was individually searched and gene expression data was manually extracted.

Data analysis: All available NB and LGG samples were downloaded, data were extracted from the Data Transfer Tool using a custom C# script [35] and processed using Microsoft Excel. In order to compare data obtained from multiple databases, we converted FPKM (Fragments Per Kilobase Million) to TPM (Transcripts Per Million) using the following equation:

$$TPM = (FPKM_g / \Sigma FPKM_s) \times 10^6$$

where $FPKM_g$ represents the FPKM of the gene of interest and $\Sigma FPKM_s$ represents the sum of all FPKM values from the patient sample [36]. Statistical analysis for the open access RNAseqV2 data was calculated using an unpaired t-test comparing NB samples to LGG.

3. Results

3.1. CNS Neuroblastoma

Among the 3,494 pediatric patients treated for CNS tumors in the Ann and Robert H. Lurie Children's Hospital of Chicago (former Children's Memorial Hospital) from September 1981 to September 2018 (37 years) only 20 cases of CNS-NB were recorded, including 12 children (0.34%) diagnosed with primary CNS-NB (all in the spinal cord) and 8 children (0.23%) diagnosed with NB metastatic to the brain (metastatic CNS-NB). Our study described 2 of our CNS-NB patients which had frozen tissue available for further analyses: (1) a primary CNS-NB that was excised from a 20 month old female patient in 1998 and was diagnosed as a NB according to the 1993 WHO classification [37] (Figure 1) and (2) a NB metastatic from a primary tumor in the adrenal gland, that was removed from a six year old female patient in 2001 and classified according to the 2000 WHO classification of CNS tumors (Figure 2) [38]. Both tumors were located at the supratentorial region of the brain.

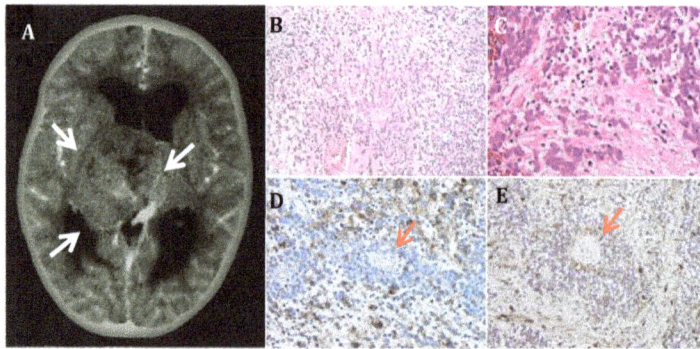

Figure 1. Primary CNS-Neuroblastoma. (**A**) Computerized Tomography image of a primary CNS-NB shows a large heterogeneous well-circumscribed lesion (arrows) measuring $5.7 \times 5.2 \times 4.8$ cm, within the right thalamus ($10\times$). (**B**,**C**). Histopathological examination shows islands of densely cellular poorly differentiated tumor cells, interspaced by sparsely cellular areas or finely fibrillary tissue. No mature neurons are identified ($10\times$ and $20\times$ respectively). (**D**) Immunostain for neuron specific enolase (NSE) ($20\times$); (**E**) Immunostain for synaptophysin ($20\times$). Homer-Wright rosettes are frequent (red arrows).

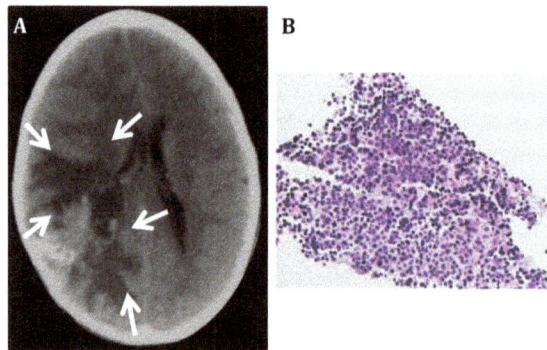

Figure 2. Neuroblastoma metastatic to the CNS. (**A**) Computerized Tomography image of a metastatic NB shows a large poorly delimited mass in the right posterior frontoparietal region of the brain (arrows). (**B**) Biopsy of the metastatic tumor mass shows small poorly differentiated cells with hyperchromatic nucleus and scant cytoplasm.

3.2. PLK4 Expression in CNS-NB Samples Determined by qRT-PCR

Three housekeeping genes (*GAPDH*, *HPRT* and *HMBS*) were used for analysis. In each individual experiment using individual housekeeping genes, CNS-NB samples showed significantly elevated *PLK4* expression levels when compared to non-embryonal brain tumors (LGG) (*GAPDH* $p = 0.0016$; *HPRT1* $p < 0.0001$; *HMBS* $p = 0.0116$) (Figure 3A–C). Accordingly, normalization of expression values using *GAPDH*, *HPRT* and *HMBS* simultaneously [26–28] also showed significant overexpression of *PLK4* in CNS-NB (FC: 15.05, $p < 0.0001$) (Figure 3D). Furthermore, in accordance with what we previously described, other embryonal brain tumor samples (ATRT and MB) also overexpressed *PLK4* ($p < 0.0001$) (Figure 3E and Table 1).

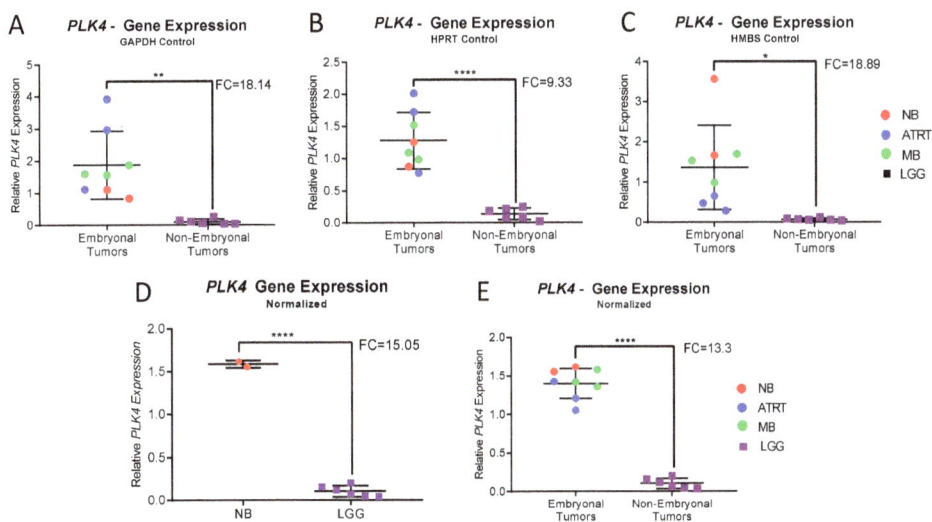

Figure 3. qRT-PCR Expression Analysis of CNS-NB, ATRT, MB and LGG. (**A–C**) Relative *PLK4* expression in CNS-NB, embryonal and non-embryonal pediatric brain tumors measured by qRT-PCR normalized to the endogenous controls *GAPDH*, *HPRT* and *HMBS* respectively, compared to LGG. (**D**) Relative *PLK4* expression in CNS-NB when normalized to all three endogenous controls compared to LGG. (**E**) Relative *PLK4* expression in embryonal tumors compared to non-embryonal tumors, normalized to all three endogenous controls. Fold changes and p-values were compared to non-embryonal pediatric brain tumors (unpaired t-tests, * $p < 0.1$, ** $p < 0.01$, **** $p < 0.0001$).

Table 1. Relative *PLK4* expression in NB, ATRT, MB and LGG. Relative *PLK4* expression measured by qRT-PCR, calculated against 3 different endogenous controls individually and normalized together.

Normalized Expression	CNS-NB	LGG	Fold Change	p-Value	Embryonal Tumors	Non-Embryonal Tumors	Fold Change	p-Value
PLK4/GAPDH	0.97	0.1	9.4	0.0016	1.88	0.1	18.14	0.006
PLK4/HPRT 1	1.062	0.14	7.78	<0.0001	1.28	0.14	9.33	0.0031
PLK4/HMBS	2.62	0.07	36.41	0.0116	1.36	0.07	18.89	0.0116

Normalized Expression	CNS-NB	LGG	Fold Change	p-Value	Embryonal Tumors	Non-Embryonal Tumors	Fold Change	p-Value
PLK4	1.58	0.1	15.05	<0.0001	1.4	0.1	13.3	<0.0001

3.3. Gene Expression Meta-Analysis

Because CNS-NB is a rare entity [39,40] and due to the limited number of samples available for molecular analysis, we performed an extensive multi-platform transcriptomic meta-analysis compiling publicly available gene expression data to validate the results observed in our patients' tumors. For this, we compared embryonal CNS tumors to non-embryonal CNS tumors represented by low grade gliomas (LGG), which is the most common form of primary CNS tumor arising in both children and adults [29,30].

The analysis of transcriptomic data from 51 normal tissue types represented in the GTEx Portal database (n = 11,688) and all NB and LGG tumor samples from the TARGET and the TCGA databases (n = 153 and 508 respectively) demonstrated that *PLK4* expression was low in almost all tissues, with 75% of them expressing ≤ 1.3 TPM (transcripts per million). The highest *PLK4* expression was observed in testis (23.7 TPM). NB showed significantly high *PLK4* expression (14.0 TPM) while LGG showed 2.2 TPM ($p < 0.0001$, unpaired t-test) (Figure 4, Tables 2 and 3).

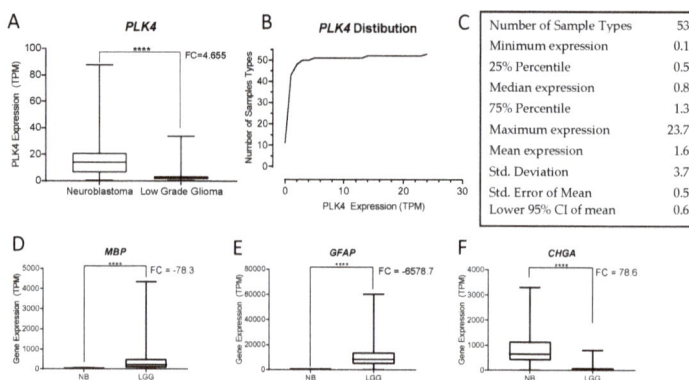

Figure 4. GDC Expression Analysis of NB and LGG. (A) *PLK4* expression in NB and LGG shows significant overexpression in neuroblastoma (NB) when compared with low grade gliomas (LGG) (**** $p < 0.0001$, unpaired *t*-test). (B) Relative frequency distribution of *PLK4* expression among samples described in Table 2: NB (TARGET $n = 153$), LGG (TCGA $n = 508$) and 51 normal human tissue sample types (GTEx Portal). (C) Descriptive statistics of *PLK4* expression in the cohort of tissue sample types (GraphPad). (D,E) The glioma markers *MBP* and *GFAP* respectively, show significant overexpression in LGG compared to NB (**** $p < 0.0001$, unpaired *t*-test). (F) The NB maker *CHGA*, shows overexpression in NB compared to LGG (**** $p < 0.0001$, unpaired *t*-test). All graphs were generated and statistics calculated using PRISM (GraphPad Software, Inc.).

Table 2. GDC and GTEx Portal Gene Expression Data. RNAseqV2 data extracted from the GDC database (Neuroblastoma and Low Grade Glioma) and GTEx Portal (51 normal human tissue samples). Genes displayed are: *PLK4* (Polo-like kinase 4), *CHGA* (Chromogranin A), *MBP* (Myelin basic protein) and *GFAP* (Glial fibrillary acidic protein). Expression is represented as median TPM (Transcripts per million) values.

Organ #	Organ Name	Sample Size	PLK4	CHGA	MBP	GFAP
	Neuroblastoma	153	14.0	658.1	2.9	0.3
	Low Grade Glioma	508	2.2	45.4	212.4	8535.2
1	Adipose—Subcutaneous	442	1.4	0.1	6.3	1.9
2	Adipose—Visceral (Omentum)	355	0.8	0.1	6.3	1.2
3	Adrenal Gland	190	0.9	7.5	1.5	0.8
4	Artery—Aorta	299	0.5	0.2	6.8	2.9
5	Artery—Coronary	173	0.8	0.2	6.5	2.1
6	Artery—Tibial	441	0.7	0.2	6.7	1.4
7	Bladder	11	1.1	0.6	7.5	0.4
8	Brain—Amygdala	100	0.6	29.2	905.8	1669.7
9	Brain—Anterior cingulate cortex (BA24)	121	0.6	87.7	302.3	1027.0
10	Brain—Caudate (Basal ganglia)	160	0.6	26.9	422.6	1577.2
11	Brain Cerebellar Hemisphere	136	0.2	4.8	208.6	600.3
12	Brain—Cerebellum	173	0.2	4.4	177.7	696.6
13	Brain—Cortex	158	0.6	219.6	267.8	1200.6
14	Brain—Frontal Cortex (BA9)	129	0.8	335.6	332.5	961.1
15	Brain—Hippocampus	123	0.4	40.5	1472.2	2225.0
16	Brain—Hypothalamus	121	0.7	79.3	890.0	3809.2
17	Brain—Nucleus accumbens (basal ganglia)	147	0.9	32.6	335.9	913.3
18	Brain—Putamen (basal ganglia)	124	0.5	23.0	884.8	985.9
19	Brain—Spinal cord (cervical c-1)	91	0.8	5.7	9405.2	12,714.4
20	Brain—Substantia nigra	88	0.5	32.2	2607.8	4370.6
21	Breast—Mammary Tissue	290	1.2	0.3	6.5	2.8
22	Cervix—Ectocervix	6	1.4	0.4	8.9	0.2
23	Cervix—Endocervix	5	1.2	1.6	11.5	0.5
24	Colon—Sigmoid	233	0.5	5.8	5.5	0.9
25	Colon—Transverse	274	1.7	38.4	5.9	0.5

Table 2. *Cont.*

Organ #	Organ Name	Sample Size	PLK4	CHGA	MBP	GFAP
26	Esophagus—Gastroesophageal Junction	244	0.5	1.9	6.1	0.8
27	Esophagus—Mucosa	407	4.5	0.2	7.1	0.3
28	Esophabus—Musclaris	370	0.5	2.2	5.4	0.6
29	Fallopian Tube	7	1.1	1.8	7.3	0.4
30	Heart—Atrial Appendage	297	0.2	0.1	2.8	2.0
31	Heart—Left Ventricle	303	0.1	0.1	2.5	1.4
32	Kidney—Cortex	45	0.5	0.4	5.1	0.6
33	Liver	175	0.2	0.1	3.6	0.3
34	Lung	427	1.2	0.3	9.9	0.9
35	Minor Salivary Gland	97	0.1	0.2	7.0	1.1
36	Muscle—Skeletal	564	0.1	0.1	9.8	0.6
37	Nerve—Tibial	414	1.2	0.3	418.9	13.3
38	Ovary	133	1.4	0.3	4.9	0.6
39	Pancreas	248	0.3	42.3	3.4	0.5
40	Pituitary	183	0.4	781.6	7.6	16.3
41	Prostate	152	0.9	7.7	6.2	0.8
42	Skin—Not Sun Exposed (Suprapubic)	387	2.7	0.4	9.9	0.9
45	Skin—Sun Exposed (Lower Leg)	473	2.7	0.4	9.5	1.0
43	Small Intestine—Terminal Ileum	137	2.0	86.5	8.1	0.4
44	Spleen	162	2.1	0.2	11.3	0.5
46	Stomach	262	0.7	226.5	5.3	0.4
47	Testis	259	23.7	158.6	3.3	1.5
48	Thyroid	446	1.0	0.3	9.4	1.4
49	Uterus	111	1.1	0.3	7.8	0.4
50	Vagina	115	2.3	0.6	7.9	0.6
51	Whole Blood	407	0.3	0.2	18.3	1.0

Table 3. Expression of *PLK4*, *CHGA*, *MBP* and *GFAP* obtained from the GDC. Genes displayed are: *PLK4* (polo-like kinase 4), *CHGA* (chromogranin A), *MBP* (myelin basic protein) and *GFAP* (glial fibrillary acidic protein). Expression is represented as average TPM (Transcripts per million) values. Statistics were calculated in PRISM (unpaired *t*-test; $p < 0.0001$).

	Neuroblastoma	Low Grade Glioma	Fold Change	*p*-Value
PLK4	14.92	3.49	4.28	$p < 0.0001$
CHGA	879.97	75.63	11.63	$p < 0.0001$
MBP	5.26	411.70	−78.32	$p < 0.0001$
GFAP	1.68	11,046.77	−6578.74	$p < 0.0001$

4. Discussion and Literature Review

Primary CNS-NB is a rare malignant embryonal tumor that can arise intracerebrally, intraorbitally or intraspinally [41]. Although it can be found in adults, it most often occurs within the first 5 years of life [39]. CNS-NB is a controversial entity which diagnostic classification has undergone a number of changes since the first publication of the WHO Classification of Tumors of the Central Nervous System in 1979 where primary NB of the CNS was classified as a "poorly differentiated neuronal tumor" [42]. In the second edition, published in 1993, CNS-NB was classified for the first time as an "embryonal tumor", which is the designation that persists today [37]. In 2000, CNS-NB was subclassified as an embryonal tumor of the "supratentorial primitive neuroectodermal tumor" (sPNET) subgroup [38]. The WHO's fourth edition classification in 2007 changed the terminology of sPNET to CNS-PNET and primary NB of the CNS was then classified as "CNS-NB" [43]. In the WHO's most recent classification published in 2016, the categorization of embryonal tumors underwent extensive changes. The term primitive neuroectodermal tumor (PNET) was eliminated from the diagnostic terminology and a category of CNS embryonal tumor "not otherwise specified", that includes tumors previously designated as CNS-PNET was created. Currently, CNS-NB is again classified as a singular entity under the umbrella of embryonal tumors [44]. Recently, extensive CNS-NB molecular data has been published. In a large study, 323 tumor samples diagnosed as CNS-PNET were subjected to histological examination, DNA methylation profiling, Affymetrix GeneChip Array and next-generation DNA and RNA sequencing analyses. Within the CNS-PNET samples, 44 CNS-NB samples were found to overexpress FOXR2 and thus categorized as "CNS NB-FOXR2" [45]. Interestingly, FOXR1 has been previously found to be overexpressed in peripheral neuroblastoma [45] and PHOX2B mutations have been recently described in NB as a potential target for therapy [46,47].

Metastases of NB to the CNS are very rare, comprising less than 10% of all cases of metastatic NB [15]. These are often osseous involving the calvarium, orbit or skull base, while primary CNS-NB commonly originate intraparenchymally spreading to the leptomeninges and subarachnoid space [48]. NB metastatic to the CNS is most commonly found within the first 18 months of age after the initial diagnosis and increased *MYCN* amplification has been reported in these recurrent tumors [41,49]. A recent meta-analysis combining profiles of NB from 761 patients with *MYCN* amplification, identified enrichment of the members of the P13K family of kinases as biomarkers of *MYCN* amplification and suggested that P13K inhibitors may represent a new therapeutic opportunity for *MYCN*-amplified NB [50].

To date, there is no established protocol for treating primary CNS-NB, references [4,51] with treatment approaches varying from palliative care to aggressive multimodality therapies. Surgery, craniospinal radiotherapy and chemotherapy have led to increased median survival, however, NB metastatic to CNS are almost universally lethal [52]. The heterogeneous nature of NB leads to diverse clinical presentations [53]. Depending on the location of the primary tumor and metastases, as well as histology and genomic data, treatment regimens range from observation of low-risk patients, to multimodal approaches in high risk patients [5]. Due to the lack of adequate drugs with sufficient brain-blood-barrier penetrance, the CNS is considered a "safe haven" for many cancer types, making both primary and metastatic CNS-NB difficult to treat and therefore, classified as high risk [52,54]. Treatments for both metastatic and primary CNS-NB begin with surgical resection, as much as possible. After surgery, a combination of chemotherapeutic agents is used, followed by craniospinal radiation and additional chemotherapy [52]. More recently GD2-targeted immunotherapy has been found to improve progression-free survival in NB metastatic to the CNS [54,55]. Stem cell implantation has been used with immunotherapy, but has not led to significantly increased survival [55].

We previously demonstrated PLK4 overexpression in pediatric embryonal brain tumors and suggested its potential as a therapeutic target for these tumors. Furthermore, it has been recently demonstrated that PLK4 is upregulated and negatively correlated with clinical outcome in peripheral NB [6]. While promising treatments for NB [54] have led to an increase in survival rates, both primary and metastatic CNS-NB have proven more difficult to treat than peripheral NB, leading to significantly decreased survival rates [4,56]. Here, we show that *PLK4* was overexpressed in CNS-NB both primary and metastatic to the CNS and validate these findings by performing a multi-platform transcriptomic meta-analysis of *PLK4* in normal and tumor tissue.

The Polo-like kinase 4 (PLK4) is a member of the polo-like family of serine/threonine protein kinases that shares little homology with the other members. While PLK1-3 have two structural polo-box domains, PLK4's second domain has been replaced with a crypto polo-box domain [15]. PLK4 is involved in cell cycle regulation and is localized to the centrosome during cell division, where it plays a major role in centriole duplication. PLK4 overexpression results in centrosome amplification, which has been found to cause genetic instability and spontaneous tumorigenesis [18,57]. PLK4 expression levels are tightly regulated by an auto-regulatory feedback loop in which PLK4 autophosphorylates its own phosphodegron, marking it for proteasomally mediated degredation [58]. This tight control maintains its expression low [59] and therefore preventing centriole over-duplication [60]. In recent years, PLK4 is becoming a subject of interest for the treatment of multiple types of adult peripheral tumors.

Although we were the first to identify PLK4 as a potential therapeutic target for pediatric embryonal tumors [7,10], PLK4 overexpression has also been described in adult peripheral tumors such as colorectal [20], breast [21], lung [22], melanoma [23], pancreatic cancer [25] and leukemia [24]. In fact, PLK4 inhibition using the small molecule CFI-400945 (CAS#1338800-06-8) [12,13,61] is currently in clinical trial for advanced solid tumors in adults (NCT01954316).

PLK4 substrates are mainly involved in cell cycle progression. PLK4 mediated phosphorylation of the centriolar assembly protein STIL, recruits STIL to site of the pre-procentriole and facilitates its interaction with Sas6 [62], which together, form the centriolar cartwheel, a complex essential to proper centriole duplication [63]. PLK4 is also involved in the regulation of centriole assembly through its direct phosphorylation of CP110, a coiled-coil protein controlling centriole length [64]. PLK4 has been found to be implicated in the localization and stabilization of the cleavage furrow through its interactions with Ect2, a Rho GEF, which activates RhoA [16]. Other notable cell cycle related PLK4 substrates include CDC25c [65], FBXW5 [66] and AURKA [67].

5. Conclusions

Our previous findings together with the findings of the present study highlight the prevalence of *PLK4* overexpression in embryonal tumors and suggest the potential of PLK4 as a new target for therapeutic intervention. Although we recognize that the number of cases in this study is small, the rarity of CNS-NB, the consistency of the results corroborated by extensive meta-analysis, the novelty and the translational potential of PLK4 as a biomarker and/or a therapeutic target is suitable for further investigation.

Author Contributions: Conceptualization: S.T.S., A.W.B., A.S. and T.T.; Study design, experiments and analyses: S.T.S., A.W.B., A.S., S.L.R. and S.G.; Pathology and image: S.T.S., P.M.C. and T.P.; Manuscript writing: S.T.S., A.W.B. and A.S.; Manuscript review and approval: S.T.S., A.W.B., A.S., P.M.C., T.P., S.L.R., S.G., and T.T.

Funding: Voices Against Brain Cancer, Musella Foundation for Cancer Research and Information.

Conflicts of Interest: The authors declare no conflict of interest.

References

1. Ramaswamy, V.; Taylor, M.D. Medulloblastoma: From myth to molecular. *J. Clin. Oncol. Off. J. Am. Soc. Clin. Oncol.* **2017**, *35*, 2355–2363. [CrossRef] [PubMed]
2. Sredni, S.T.; Tomita, T. Rhabdoid tumor predisposition syndrome. *Pediatr. Dev. Pathol.* **2015**, *18*, 49–58. [CrossRef] [PubMed]
3. Tariq, M.U.; Ahmad, Z.; Minhas, M.K.; Memon, A.; Mushtaq, N.; Hawkins, C. Embryonal tumor with multilayered rosettes, c19mc-altered: Report of an extremely rare malignant pediatric central nervous system neoplasm. *SAGE Open Med. Case Rep.* **2017**, *5*, 2050313X17745208. [CrossRef] [PubMed]
4. Bianchi, F.; Tamburrini, G.; Gessi, M.; Frassanito, P.; Massimi, L.; Caldarelli, M. Central nervous system (cns) neuroblastoma. A case-based update. *Child. Nerv. Syst.* **2018**, *34*, 817–823. [CrossRef] [PubMed]
5. Maris, J.M.; Hogarty, M.D.; Bagatell, R.; Cohn, S.L. Neuroblastoma. *Lancet (London, England)* **2007**, *369*, 2106–2120. [CrossRef]
6. Tian, X.; Zhou, D.; Chen, L.; Tian, Y.; Zhong, B.; Cao, Y.; Dong, Q.; Zhou, M.; Yan, J.; Wang, Y.; et al. Polo-like kinase 4 mediates epithelial-mesenchymal transition in neuroblastoma via pi3k/akt signaling pathway. *Cell Death Dis.* **2018**, *9*, 54. [CrossRef] [PubMed]
7. Sredni, S.T.; Suzuki, M.; Yang, J.P.; Topczewski, J.; Bailey, A.W.; Gokirmak, T.; Gross, J.N.; de Andrade, A.; Kondo, A.; Piper, D.R.; et al. A functional screening of the kinome identifies the polo-like kinase 4 as a potential therapeutic target for malignant rhabdoid tumors, and possibly, other embryonal tumors of the brain. *Pediatr. Blood Cancer* **2017**, *64*, e26551. [CrossRef] [PubMed]
8. Zhang, Z.K.; Davies, K.P.; Allen, J.; Zhu, L.; Pestell, R.G.; Zagzag, D.; Kalpana, G.V. Cell cycle arrest and repression of cyclin d1 transcription by ini1/hsnf5. *Mol. Cell. Biol.* **2002**, *22*, 5975–5988. [CrossRef] [PubMed]
9. Albanese, P.; Belin, M.F.; Delattre, O. The tumour suppressor hsnf5/ini1 controls the differentiation potential of malignant rhabdoid cells. *Eur. J. Cancer (Oxford, England: 1990)* **2006**, *42*, 2326–2334. [CrossRef] [PubMed]

10. Sredni, S.T.; Bailey, A.W.; Suri, A.; Hashizume, R.; He, X.; Louis, N.; Gokirmak, T.; Piper, D.R.; Watterson, D.M.; Tomita, T. Inhibition of polo-like kinase 4 (plk4): A new therapeutic option for rhabdoid tumors and pediatric medulloblastoma. *Oncotarget* **2017**, *8*, 111190–111212. [CrossRef] [PubMed]

11. Sredni, S.T.; Tomita, T. The polo-like kinase 4 gene (plk4) is overexpressed in pediatric medulloblastoma. *Child. Nerv. Syst.* **2017**, *33*, 1031. [CrossRef] [PubMed]

12. Sampson, P.B.; Liu, Y.; Forrest, B.; Cumming, G.; Li, S.W.; Patel, N.K.; Edwards, L.; Laufer, R.; Feher, M.; Ban, F.; et al. The discovery of polo-like kinase 4 inhibitors: Identification of (1r,2s).2-(3-((e).4-(((cis).2,6-dimethylmorpholino)methyl)styryl). 1h.Indazol-6-yl)-5′-methoxyspiro [cyclopropane-1,3′-indolin]-2′-one (cfi-400945) as a potent, orally active antitumor agent. *J. Med. Chem.* **2015**, *58*, 147–169. [CrossRef] [PubMed]

13. Mason, J.M.; Lin, D.C.; Wei, X.; Che, Y.; Yao, Y.; Kiarash, R.; Cescon, D.W.; Fletcher, G.C.; Awrey, D.E.; Bray, M.R.; et al. Functional characterization of cfi-400945, a polo-like kinase 4 inhibitor, as a potential anticancer agent. *Cancer Cell* **2014**, *26*, 163–176. [CrossRef] [PubMed]

14. Yu, B.; Yu, Z.; Qi, P.P.; Yu, D.Q.; Liu, H.M. Discovery of orally active anticancer candidate cfi-400945 derived from biologically promising spirooxindoles: Success and challenges. *Eur. J. Med. Chem.* **2015**, *95*, 35–40. [CrossRef] [PubMed]

15. Sillibourne, J.E.; Bornens, M. Polo-like kinase 4: The odd one out of the family. *Cell Division* **2010**, *5*, 25. [CrossRef] [PubMed]

16. Rosario, C.O.; Kazazian, K.; Zih, F.S.; Brashavitskaya, O.; Haffani, Y.; Xu, R.S.; George, A.; Dennis, J.W.; Swallow, C.J. A novel role for plk4 in regulating cell spreading and motility. *Oncogene* **2015**, *34*, 3441–3451. [CrossRef] [PubMed]

17. Bettencourt-Dias, M.; Rodrigues-Martins, A.; Carpenter, L.; Riparbelli, M.; Lehmann, L.; Gatt, M.K.; Carmo, N.; Balloux, F.; Callaini, G.; Glover, D.M. Sak/plk is required for centriole duplication and flagella development. *Curr. Biol. CB* **2005**, *15*, 2199–2207. [CrossRef] [PubMed]

18. Levine, M.S.; Bakker, B.; Boeckx, B.; Moyett, J.; Lu, J.; Vitre, B.; Spierings, D.C.; Lansdorp, P.M.; Cleveland, D.W.; Lambrechts, D.; et al. Centrosome amplification is sufficient to promote spontaneous tumorigenesis in mammals. *Dev. Cell* **2017**, *40*, 313–322. [CrossRef] [PubMed]

19. Shinmura, K.; Kurabe, N.; Goto, M.; Yamada, H.; Natsume, H.; Konno, H.; Sugimura, H. Plk4 overexpression and its effect on centrosome regulation and chromosome stability in human gastric cancer. *Mol. Biol. Rep.* **2014**, *41*, 6635–6644. [CrossRef] [PubMed]

20. Macmillan, J.C.; Hudson, J.W.; Bull, S.; Dennis, J.W.; Swallow, C.J. Comparative expression of the mitotic regulators sak and plk in colorectal cancer. *Ann. Surg. Oncol.* **2001**, *8*, 729–740. [CrossRef] [PubMed]

21. Marina, M.; Saavedra, H.I. Nek2 and plk4: Prognostic markers, drivers of breast tumorigenesis and drug resistance. *Front. Biosci. (Landmark Edition)* **2014**, *19*, 352–365. [CrossRef]

22. Kawakami, M.; Mustachio, L.M.; Zheng, L.; Chen, Y.; Rodriguez-Canales, J.; Mino, B.; Kurie, J.M.; Roszik, J.; Villalobos, P.A.; Thu, K.L.; et al. Polo-like kinase 4 inhibition produces polyploidy and apoptotic death of lung cancers. *Proc. Natl. Acad. Sci. USA* **2018**, *115*, 1913–1918. [CrossRef] [PubMed]

23. Denu, R.A.; Shabbir, M.; Nihal, M.; Singh, C.K.; Longley, B.J.; Burkard, M.E.; Ahmad, N. Centriole overduplication is the predominant mechanism leading to centrosome amplification in melanoma. *Mol. Cancer Res. MCR* **2018**, *16*, 517–527. [CrossRef] [PubMed]

24. Goroshchuk, O.; Kolosenko, I.; Vidarsdottir, L.; Azimi, A.; Palm-Apergi, C. Polo-like kinases and acute leukemia. *Oncogene* **2018**. [CrossRef] [PubMed]

25. Lohse, I.; Mason, J.; Cao, P.M.; Pintilie, M.; Bray, M.; Hedley, D.W. Activity of the novel polo-like kinase 4 inhibitor cfi-400945 in pancreatic cancer patient-derived xenografts. *Oncotarget* **2017**, *8*, 3064–3071. [CrossRef] [PubMed]

26. Vandesompele, J.; De Preter, K.; Pattyn, F.; Poppe, B.; Van Roy, N.; De Paepe, A.; Speleman, F. Accurate normalization of real-time quantitative rt-pcr data by geometric averaging of multiple internal control genes. *Genome Biol.* **2002**, *3*, Research0034. [CrossRef] [PubMed]

27. Haller, F.; Kulle, B.; Schwager, S.; Gunawan, B.; von Heydebreck, A.; Sultmann, H.; Fuzesi, L. Equivalence test in quantitative reverse transcription polymerase chain reaction: Confirmation of reference genes suitable for normalization. *Anal. Biochem.* **2004**, *335*, 1–9. [CrossRef] [PubMed]

28. Valente, V.; Teixeira, S.A.; Neder, L.; Okamoto, O.K.; Oba-Shinjo, S.M.; Marie, S.K.; Scrideli, C.A.; Paco-Larson, M.L.; Carlotti, C.G., Jr. Selection of suitable housekeeping genes for expression analysis in glioblastoma using quantitative rt-pcr. *Ann. Neurosci.* **2014**, *21*, 62–63. [CrossRef] [PubMed]

29. Goodenberger, M.L.; Jenkins, R.B. Genetics of adult glioma. *Cancer Genet.* **2012**, *205*, 613–621. [CrossRef] [PubMed]

30. Packer, R.J.; Pfister, S.; Bouffet, E.; Avery, R.; Bandopadhayay, P.; Bornhorst, M.; Bowers, D.C.; Ellison, D.; Fangusaro, J.; Foreman, N.; et al. Pediatric low-grade gliomas: Implications of the biologic era. *Neuro-Oncol.* **2017**, *19*, 750–761. [CrossRef] [PubMed]

31. Georgantzi, K.; Sköldenberg, E.G.; Stridsberg, M.; Kogner, P.; Jakobson, Å.; Janson, E.T.; Christofferson, R.H.B. Chromogranin a and neuron-specific enolase in neuroblastoma: Correlation to stage and prognostic factors. *Pediatr. Hematol. Oncol.* **2018**, *35*, 156–165. [CrossRef] [PubMed]

32. Popko, B.; Pearl, D.K.; Walker, D.M.; Comas, T.C.; Baerwald, K.D.; Burger, P.C.; Scheithauer, B.W.; Yates, A.J. Molecular markers that identify human astrocytomas and oligodendrogliomas. *J. Neuropathol. Exp. Neurol.* **2002**, *61*, 329–338. [CrossRef] [PubMed]

33. Hol, E.M.; Pekny, M. Glial fibrillary acidic protein (gfap) and the astrocyte intermediate filament system in diseases of the central nervous system. *Curr. Opin. Cell Biol.* **2015**, *32*, 121–130. [CrossRef] [PubMed]

34. Consortium, G.T.; Aguet, F.; Brown, A.A.; Castel, S.E.; Davis, J.R.; He, Y.; Jo, B.; Mohammadi, P.; Park, Y.; Parsana, P.; et al. Genetic effects on gene expression across human tissues. *Nature* **2017**, *550*, 204.

35. Siddiqui, S.; White, M.W.; Schroeder, A.M.; DeLuca, N.V.; Leszczynski, A.L.; Raimondi, S.L. Aberrant dnmt3b7 expression correlates to tissue type, stage, and survival across cancers. *PLoS ONE* **2018**, *13*, e0201522. [CrossRef] [PubMed]

36. Loir, P. Models for transcript quantification for rna-seq. *arXiv* **2011**.

37. Kleihues, P.; Burger, P.C.; Scheithauer, B.W. The new who classification of brain tumours. *Brain Pathol. (Zurich, Switzerland)* **1993**, *3*, 255–268. [CrossRef] [PubMed]

38. Kleihues, P.; Louis, D.N.; Scheithauer, B.W.; Rorke, L.B.; Reifenberger, G.; Burger, P.C.; Cavenee, W.K. The who classification of tumors of the nervous system. *J. Neuropathol. Exp. Neurol.* **2002**, *61*, 215–225. [CrossRef] [PubMed]

39. Horten, B.C.; Rubinstein, L.J. Primary cerebral neuroblastoma. A clinicopathological study of 35 cases. *Brain A J. Neurol.* **1976**, *99*, 735–756. [CrossRef]

40. Etus, V.; Kurtkaya, O.; Sav, A.; Ilbay, K.; Ceylan, S. Primary cerebral neuroblastoma: A case report and review. *Tohoku J. Exp. Med.* **2002**, *197*, 55–65. [CrossRef] [PubMed]

41. Latchaw, R.E.; L'Heureux, P.R.; Young, G.; Priest, J.R. Neuroblastoma presenting as central nervous system disease. *AJNR. Am. J. Neuroradiol.* **1982**, *3*, 623–630. [PubMed]

42. Zulch, K.J. Principles of the new world health organization (who) classification of brain tumors. *Neuroradiology* **1980**, *19*, 59–66. [CrossRef] [PubMed]

43. Louis, D.N.; Ohgaki, H.; Wiestler, O.D.; Cavenee, W.K.; Burger, P.C.; Jouvet, A.; Scheithauer, B.W.; Kleihues, P. The 2007 who classification of tumours of the central nervous system. *Acta Neuropathol.* **2007**, *114*, 97–109. [CrossRef] [PubMed]

44. Louis, D.N.; Perry, A.; Reifenberger, G.; von Deimling, A.; Figarella-Branger, D.; Cavenee, W.K.; Ohgaki, H.; Wiestler, O.D.; Kleihues, P.; Ellison, D.W. The 2016 world health organization classification of tumors of the central nervous system: A summary. *Acta Neuropathol.* **2016**, *131*, 803–820. [CrossRef] [PubMed]

45. Sturm, D.; Orr, B.A.; Toprak, U.H.; Hovestadt, V.; Jones, D.T.W.; Capper, D.; Sill, M.; Buchhalter, I.; Northcott, P.A.; Leis, I.; et al. New brain tumor entities emerge from molecular classification of cns-pnets. *Cell* **2016**, *164*, 1060–1072. [CrossRef] [PubMed]

46. Alexandrescu, S.; Paulson, V.; Dubuc, A.; Ligon, A.; Lidov, H.G. Phox2b is a reliable immunomarker in distinguishing peripheral neuroblastic tumours from cns embryonal tumours. *Histopathology* **2018**. [CrossRef] [PubMed]

47. Cardani, S.; Di Lascio, S.; Belperio, D.; Di Biase, E.; Ceccherini, I.; Benfante, R.; Fornasari, D. Desogestrel down-regulates phox2b and its target genes in progesterone responsive neuroblastoma cells. *Exp. Cell Res.* **2018**, *370*, 671–679. [CrossRef] [PubMed]

48. Zimmerman, R.A.; Bilaniuk, L.T. Ct of primary and secondary craniocerebral neuroblastoma. *Am. J. Roentgenol.* **1980**, *135*, 1239–1242. [CrossRef] [PubMed]

49. Matthay, K.K.; Brisse, H.; Couanet, D.; Couturier, J.; Benard, J.; Mosseri, V.; Edeline, V.; Lumbroso, J.; Valteau-Couanet, D.; Michon, J. Central nervous system metastases in neuroblastoma: Radiologic, clinical, and biologic features in 23 patients. *Cancer* **2003**, *98*, 155–165. [CrossRef] [PubMed]

50. Petrov, I.; Suntsova, M.; Ilnitskaya, E.; Roumiantsev, S.; Sorokin, M.; Garazha, A.; Spirin, P.; Lebedev, T.; Gaifullin, N.; Larin, S.; et al. Gene expression and molecular pathway activation signatures of mycn-amplified neuroblastomas. *Oncotarget* **2017**, *8*, 83768–83780. [CrossRef] [PubMed]

51. Mishra, A.; Beniwal, M.; Nandeesh, B.N.; Srinivas, D.; Somanna, S. Primary pediatric intracranial neuroblastoma: A report of two cases. *J Pediatr. Neurosci.* **2018**, *13*, 366–370. [PubMed]

52. Kramer, K.; Kushner, B.; Heller, G.; Cheung, N.K. Neuroblastoma metastatic to the central nervous system. The memorial sloan-kettering cancer center experience and a literature review. *Cancer* **2001**, *91*, 1510–1519. [CrossRef]

53. Kholodenko, I.V.; Kalinovsky, D.V.; Doronin, I.I.; Deyev, S.M.; Kholodenko, R.V. Neuroblastoma origin and therapeutic targets for immunotherapy. *J. Immunol. Res.* **2018**, *2018*, 7394268. [CrossRef] [PubMed]

54. Matthay, K.K.; Maris, J.M.; Schleiermacher, G.; Nakagawara, A.; Mackall, C.L.; Diller, L.; Weiss, W.A. Neuroblastoma. *Nat. Rev. Dis. Primers* **2016**, *2*, 16078. [CrossRef] [PubMed]

55. Kushner, B.H.; Ostrovnaya, I.; Cheung, I.Y.; Kuk, D.; Kramer, K.; Modak, S.; Yataghene, K.; Cheung, N.-K.V. Prolonged progression-free survival after consolidating second or later remissions of neuroblastoma with anti-gd2 immunotherapy and isotretinoin: A prospective phase ii study. *OncoImmunology* **2015**, *4*, e1016704. [CrossRef] [PubMed]

56. Kramer, K.; Kushner, B.H.; Modak, S.; Pandit-Taskar, N.; Smith-Jones, P.; Zanzonico, P.; Humm, J.L.; Xu, H.; Wolden, S.L.; Souweidane, M.M.; et al. Compartmental intrathecal radioimmunotherapy: Results for treatment for metastatic cns neuroblastoma. *J. Neuro-Oncol.* **2010**, *97*, 409–418. [CrossRef] [PubMed]

57. Ko, M.A.; Rosario, C.O.; Hudson, J.W.; Kulkarni, S.; Pollett, A.; Dennis, J.W.; Swallow, C.J. Plk4 haploinsufficiency causes mitotic infidelity and carcinogenesis. *Nat. Genet.* **2005**, *37*, 883–888. [CrossRef] [PubMed]

58. Holland, A.J.; Lan, W.; Niessen, S.; Hoover, H.; Cleveland, D.W. Polo-like kinase 4 kinase activity limits centrosome overduplication by autoregulating its own stability. *J. Cell Biol.* **2010**, *188*, 191–198. [CrossRef] [PubMed]

59. Fode, C.; Binkert, C.; Dennis, J.W. Constitutive expression of murine sak-a suppresses cell growth and induces multinucleation. *Mol. Cell. Biol.* **1996**, *16*, 4665–4672. [CrossRef] [PubMed]

60. Sillibourne, J.E.; Tack, F.; Vloemans, N.; Boeckx, A.; Thambirajah, S.; Bonnet, P.; Ramaekers, F.C.; Bornens, M.; Grand-Perret, T. Autophosphorylation of polo-like kinase 4 and its role in centriole duplication. *Mol. Biol. Cell* **2010**, *21*, 547–561. [CrossRef] [PubMed]

61. Sampson, P.B.; Liu, Y.; Patel, N.K.; Feher, M.; Forrest, B.; Li, S.W.; Edwards, L.; Laufer, R.; Lang, Y.; Ban, F.; et al. The discovery of polo-like kinase 4 inhibitors: Design and optimization of spiro[cyclopropane-1,3'[3h]indol]-2'(1'h).Ones as orally bioavailable antitumor agents. *J. Med. Chem.* **2015**, *58*, 130–146. [CrossRef] [PubMed]

62. Dzhindzhev, N.S.; Tzolovsky, G.; Lipinszki, Z.; Abdelaziz, M.; Debski, J.; Dadlez, M.; Glover, D.M. Two-step phosphorylation of ana2 by plk4 is required for the sequential loading of ana2 and sas6 to initiate procentriole formation. *Open Biol.* **2017**, *7*, 170247. [CrossRef] [PubMed]

63. Kim, M.; O'Rourke, B.P.; Soni, R.K.; Jallepalli, P.V.; Hendrickson, R.C.; Tsou, M.B. Promotion and suppression of centriole duplication are catalytically coupled through plk4 to ensure centriole homeostasis. *Cell Rep.* **2016**, *16*, 1195–1203. [CrossRef] [PubMed]

64. Lee, M.; Seo, M.Y.; Chang, J.; Hwang, D.S.; Rhee, K. Plk4 phosphorylation of cp110 is required for efficient centriole assembly. *Cell Cycle* **2017**, *16*, 1225–1234. [CrossRef] [PubMed]

65. Bonni, S.; Ganuelas, M.L.; Petrinac, S.; Hudson, J.W. Human plk4 phosphorylates cdc25c. *Cell Cycle* **2008**, *7*, 545–547. [CrossRef] [PubMed]

66. Puklowski, A.; Homsi, Y.; Keller, D.; May, M.; Chauhan, S.; Kossatz, U.; Grunwald, V.; Kubicka, S.; Pich, A.;
     Manns, M.P.; et al. The scf-fbxw5 e3-ubiquitin ligase is regulated by plk4 and targets hssas-6 to control
     centrosome duplication. *Nat. Cell Biol.* **2011**, *13*, 1004–1009. [CrossRef] [PubMed]
67. Bury, L.; Coelho, P.A.; Simeone, A.; Ferries, S.; Eyers, C.E.; Eyers, P.A.; Zernicka-Goetz, M.; Glover, D.M. Plk4
     and aurora a cooperate in the initiation of acentriolar spindle assembly in mammalian oocytes. *J. Cell Biol.*
     2017, *216*, 3571–3590. [CrossRef] [PubMed]

© 2018 by the authors. Licensee MDPI, Basel, Switzerland. This article is an open access
article distributed under the terms and conditions of the Creative Commons Attribution
(CC BY) license (http://creativecommons.org/licenses/by/4.0/).

![bioengineering logo] *bioengineering*

MDPI

Review

The Emerging Role of Amino Acid PET in Neuro-Oncology

Amer M. Najjar [1,*], Jason M. Johnson [2] and Dawid Schellingerhout [2]

[1] Division of Pediatrics, The University of Texas M.D. Anderson Cancer Center, Houston, TX 77030, USA
[2] Department of Diagnostic Radiology—Neuro Imaging, The University of Texas M.D. Anderson Cancer Center, Houston, TX 77030, USA; jjohnson12@mdanderson.org (J.M.J.); dawid.schellingerhout@mdanderson.org (D.S.)
*   Correspondence: amer.najjar@mdanderson.org

Received: 15 September 2018; Accepted: 21 November 2018; Published: 28 November 2018

Abstract: Imaging plays a critical role in the management of the highly complex and widely diverse central nervous system (CNS) malignancies in providing an accurate diagnosis, treatment planning, response assessment, prognosis, and surveillance. Contrast-enhanced magnetic resonance imaging (MRI) is the primary modality for CNS disease management due to its high contrast resolution, reasonable spatial resolution, and relatively low cost and risk. However, defining tumor response to radiation treatment and chemotherapy by contrast-enhanced MRI is often difficult due to various factors that can influence contrast agent distribution and perfusion, such as edema, necrosis, vascular alterations, and inflammation, leading to pseudoprogression and pseudoresponse assessments. Amino acid positron emission tomography (PET) is emerging as the method of resolving such equivocal lesion interpretations. Amino acid radiotracers can more specifically differentiate true tumor boundaries from equivocal lesions based on their specific and active uptake by the highly metabolic cellular component of CNS tumors. These therapy-induced metabolic changes detected by amino acid PET facilitate early treatment response assessments. Integrating amino acid PET in the management of CNS malignancies to complement MRI will significantly improve early therapy response assessment, treatment planning, and clinical trial design.

Keywords: magnetic resonance imaging; positron emission tomography; amino acid PET; central nervous system malignancy; pseudoprogression; pseudoresponse

1. Introduction

Malignancies of the central nervous system (CNS) account for an estimated 23,000 cases and over 16,000 deaths each year [1]. Cerebral gliomas are second to meningiomas in frequency and account for the highest number of cancer mortalities in adults under the age of 35 [2]. Brain tumors arising from metastasis originating from peripheral tumors such as lymphoma, melanoma, lung, and breast cancer occur at an even a higher rate with an incidence of 9–17% [3].

The most recent 2016 World Health Organization (WHO) Classification of Tumors of the Central Nervous System defines CNS malignancies within four categories (grades I, II, III, and IV) based on molecular parameters and histology [4]. Grades I and II gliomas are devoid of anaplastic features and are classified as low-grade gliomas (LGG). High-grade gliomas (HGG) are classified in grades III and IV and include the most aggressive form, glioblastoma (grade IV), which has a median overall survival of 1.5 years [5].

The complexity and diversity of CNS malignancies necessitate a multifaceted approach to therapy that includes surgery, radiation treatment, chemotherapy and, more recently, immunotherapy. Historically, therapy of CNS tumors entailed surgery and radiotherapy. Improving outcomes relied on radiotherapy dose escalation and responses were measured by overall survival. With the advent of chemotherapeutics and immunotherapy (bevacizumab), radiographic assessment became necessary to assess immediate responses manifested in anatomic and molecular changes. Thus, imaging became an integral component of every stage of CNS disease management providing information that is critical to staging, formulating preoperative strategies, monitoring therapy response, surveillance, and prognosis.

2. Role of Vascularity in Magnetic Resonance Imaging of CNS Tumors

Magnetic resonance imaging (MRI) is the primary diagnostic method given its high soft tissue contrast, spatial resolution, low risk, ready availability and relatively low cost. Intensity contrast between tumor mass and surrounding brain tissue along with anatomical distortions of normal brain structures and contrast-enhanced regions typically delineate tumors in MR images although the tumor boundaries are often notoriously difficult to demonstrate accurately by imaging. T1- and T2-weighted and fluid-attenuated inversion recovery (FLAIR) are the standard sequences utilized. Compared to healthy brain tissue, CNS tumors typically appear hypointense to myelinated white matter on T1-weighted images and hyperintense on T2. Other structural characteristics associated with tumor mass may include cysts, necrosis, hemorrhage, and calcification.

CNS tumors are generally hyper-vascularized in contrast to the highly structured and selectively permeable blood-brain barrier (BBB), which acts to protect the privileged chemical environment of the brain. The BBB is formed by tight junctions between endothelial cells supported by pericytes and astrocytic foot processes limiting permeability to the vast majority of circulating agents [6]. Although microglia within the vicinity of blood vessels can repair a transient injury to the BBB [7], pathologies of the brain can compromise the integrity of the barrier increasing permeability to large therapeutic agents such as antibody drugs [8–10]. Access of anti-CTLA antibodies and bevacizumab, for example, depend explicitly on a compromise of the BBB [11].

In contrast to the highly integrated nature of the BBB [12,13], tumor vascularity is irregular, leaky, and poorly structured. This abnormality gives rise to permeability, interstitial fluid pressure, hypoxia, necrosis, and edema characteristically exhibited by glioblastomas. Aberrant tumor vascularity can be established within the structured BBB environment by metastatic cancer cells that are capable of breaching and penetrating the tight junctions of the barrier through adherence and proteolytic processes that mimic leukocyte extravasation. Once established beyond the BBB, the metastatic tumor microenvironment signals the development of a new heterogenic vascular supply characterized by increased permeability due to altered pericyte composition [14]. This leaky neovasculature can favorably influence the delivery of therapeutics by way of leakiness but may also unfavorably raise interstitial pressures to resist penetration of therapeutic agents [12].

The vascular disparity between the tumor and healthy brain tissue facilitates contrast-enhanced tumor resolution with gadolinium agents for critical response assessments of CNS tumors by MRI [15–17]. Increased tumor vascularity is a surrogate of elevated proliferation and aggressiveness and has been employed to delineate tumors through perfusion contrast-enhancing agents to diagnose and monitor brain tumor response. Congruently, the hypervascularity of glioblastomas has also been exploited as a therapeutic target of antiangiogenic drugs such as bevacizumab. Although bevacizumab has yielded little improvement of overall survival, an undefined subset of glioma patients do receive survival benefit from this agent, and a marginal improvement of progression-free survival and quality of life has justified the use of the drug in combination with standard-of-care regimens [18].

3. Limitations of Treatment Response Assessments by MRI

The role of imaging in CNS tumor management has evolved to meet the needs of advances in therapy. Early therapeutic approaches relied primarily on resection and postoperative radiotherapy of CNS malignancies. These measures provided a survival benefit which was reflected in overall survival as the primary endpoint. Renewed efforts over the last 30 years to improve therapeutic outcomes through radiation dose escalation and adjuvant chemotherapy necessitated formulating new early response assessment parameters to provide objective and mechanistic insights into treatment response [19]. Consequently, the Macdonald criteria were established in 1990 as a means of reporting early radiographic response based on contrast-enhanced computed tomography and MRI [20]. The Macdonald criteria use contrast enhancement metrics to objectively stratify therapeutic responses into four categories: (1) complete response, (2) partial response, (3) stable disease, and (4) progressive disease [20,21].

Progression-free survival is a more immediate assessment of therapeutic efficacy and serves the needs of a more precise readout for specific therapies. This requires an accurate imaging readout at chosen intermediate time points during therapy. However, the limitations of anatomic and volumetric measurements, compounded by the subjective interpretation of equivocal lesions have hindered the ability of current imaging modalities in providing an intermediate clinical readout. These inadequacies with the Macdonald metrics became apparent with recognition of contrast-enhancing or -diminishing artifacts elicited by radiation-induced necrosis and the alteration of vascularity by chemotherapy (temozolomide) and immunotherapies (bevacizumab) leading to misinterpretation of therapeutic responses. These artifacts have introduced new caveats in the interpretation of radiological data based on the Macdonald criteria.

Therapies affecting vascular permeability and perfusion give rise to the phenomena of pseudoprogression and pseudoresponse where the tumor alternately appears worse or better on imaging due to spurious effects on the vasculature. This presents a formidable challenge to the accurate and objective evaluation of therapeutic outcomes, as intermediate time points gauging progression-free survival become very difficult to interpret (Figure 1).

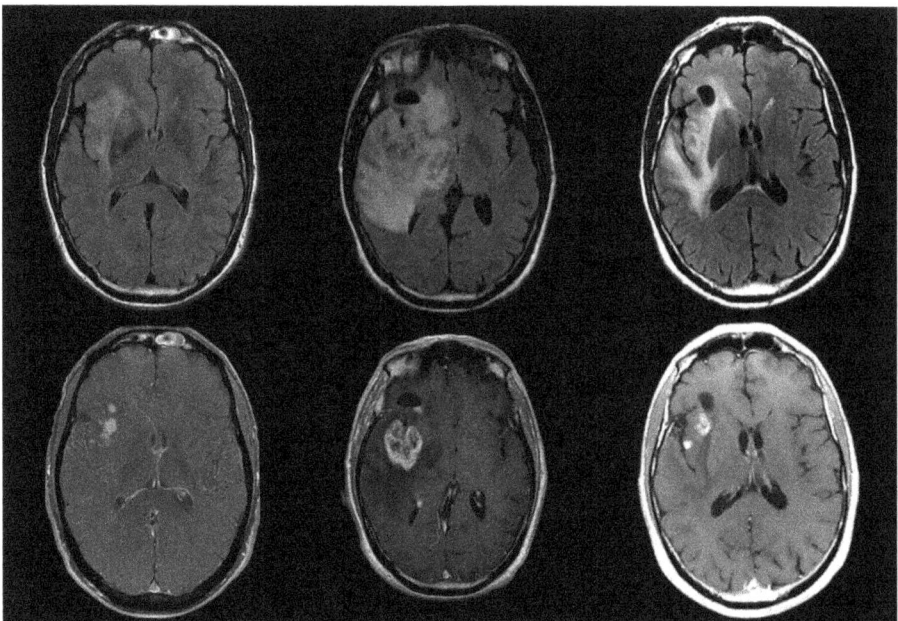

Figure 1. Glioblastoma (WHO IV) patient at presentation (left), shows an insular tumor with islands of enhancement. Following 60 Gy of radiation with Temozolamide 75 mg/m^2 daily (middle column), there is an apparent increase in both enhancement and edema with mass effect. These worsened imaging findings resolve one month later with Decadron 6 mg twice daily, with a near return to imaging baseline. This apparent worsening on imaging is known as pseudoprogression and represents an inflammatory response to therapy that is difficult to distinguish from true progression. Pseudoprogression complicates the imaging and clinical management of glioma patients.

Increased BBB permeability, necrosis, inflammation, and hemorrhage may be instigated by radiation-induced injury along with edema, and can appear mass-like on imaging. Unlike a tumor mass, however, these contrast-enhancing regions are not associated with cellular density or vascular intensity but represent a site of tissue breakdown and leakiness that mimics many of the imaging attributes of a tumor. The combination of radiotherapy and cytotoxic agents will often cause this effect and is known as pseudoprogression, which mimics the imaging appearance of tumor progression and can even cause clinical symptoms due to mass effect but is not due to true progressive disease. Checkpoint-blockade immunotherapy is also likely to present as pseudoprogression [22].

Conversely, antiangiogenic agents, such as bevacizumab, may cause short-term decreased perfusion and reduced contrast enhancement due to "normalization" of the tumor vasculature leading to a false appearance of treatment response, or pseudoresponse, not associated with a real anti-tumor effect or improved overall survival [23,24]. Bevacizumab's effects can cause imaging to show decreased T1 contrast enhancement, edema, and mass effect. This imaging outcome, however, does not correlate with long-term benefit or improved overall survival [24].

The challenges and limitations associated with MRI limit objective assessment of timely therapy response assessment and prognostication. Contrast-enhanced MRI is mostly a function of BBB integrity and tumor vascularity and is, therefore, a nonspecific form of tumor mass characterization that is prone to equivocal interpretation. These imaging artifacts may be instigated by transient effects of therapeutic interventions enhancing contrast leading to overestimation of their therapeutic efficacy. Therefore, evaluation over multiple time points is critical to circumvent misinterpretation of the single-point transient anomaly.

4. Positron Emission Tomography of CNS Tumors

Positron emission tomography (PET) is an imaging modality that is based on the preferential uptake and retention of radiolabeled tracers by the target tissue. These radiotracers mimic or are sometimes chemically identical to, metabolites that are avidly taken up by proliferative cells to meet their energy or biomass demands. Tumor cells have a higher tendency to absorb these metabolites generating contrast in uptake between tumor mass and surrounding healthy tissue.

The development of PET radiotracers has addressed some of the limitations of structural MRI in discerning pseudoprogression from true progression. Tumor delineation based on metabolic radiotracer uptake offers a functional basis of detection that yields enhanced differentiation of tumor from equivocal lesions over MRI, improved delineation of tumor boundaries for surgery and radiotherapy planning, differentiation between tumor progression and treatment-related responses, and monitoring of tumor response to therapy.

Over the last four decades, PET with 2-deoxy-2-[^{18}F]-fluoro-D-glucose (^{18}F-FDG PET) combined with computed tomography has emerged as a standard-of-care imaging modality for the detection of tumors based on their elevated glucose metabolic rate (Warburg effect). The utility of ^{18}F-FDG PET in imaging brain tumors, however, is hindered by the elevated background uptake level of glucose by normal brain tissue generating little discernable contrast. Moreover, immune cell activation within inflammatory responses also exhibits elevated glucose metabolism further obscuring distinction from brain or tumor tissue [25]. While metabolic contrast between tumor and healthy brain for FDG is low, there are considerable differences for amino acid and nucleotide metabolism. Tumor cell division (DNA replication) and growth (biomass generation) demand amino acid and nucleotide building blocks to meet the needs of the rapidly proliferating cells.

Accordingly, elevated DNA replication of tumor cells can be readily discerned using 3'-deoxy-3'-^{18}F-fluorothymidine (^{18}F-FLT) PET against the background of the relatively quiescent normal brain cells. Importantly, ^{18}F-FLT does not readily accumulate in inflammatory lesions, as is the case with ^{18}F-FDG, and is, therefore, able to report on cellular proliferation. Inflammatory immune cells primarily undergo aerobic glycolysis during activation, and are, therefore, not detected by ^{18}F-FLT. The specificity of ^{18}F-FLT for cellular proliferation facilitates its utility in reporting on tumor response to therapy [26] and has been shown to a better predictor of overall survival than MRI following treatment of recurrent gliomas with bevacizumab and irinotecan [27]. Moreover, ^{18}F-FLT uptake detects treatment response early and is highly predictive of overall and progression-free survival following bevacizumab treatment [28,29]. However, the need for BBB breakdown for ^{18}F-FLT limits its reliability in reporting on brain tumor treatment response [30].

4.1. Amino Acid PET

Amino acid PET has been evaluated extensively for detecting tumor mass based on metabolic tumor volume and is playing an increasingly important role in the management of CNS tumors. Ambiguous brain lesions that may complicate an accurate diagnosis of brain malignancies include hemorrhage, necrosis, edema, infarctions, abscesses, and inflammation. These extraneous lesions may be caused by a response to radio- and chemotherapy (inflammation and necrosis) or secondary anatomical disruptions caused by tumor outgrowth. In contrast to structural MRI, which cannot accurately differentiate lesion traits, amino acid radiotracer accumulation is a function of tumor avidity for the carbon source to meet its high demands for biomass and energy generation. This differential uptake of amino acid tracers can be exploited to specifically delineate cellular mass and tumor boundaries from surrounding normal tissue. Analogs of methionine, tyrosine, phenylalanine, alanine, and leucine have been evaluated for their specificity in delineating CNS tumors yielding nearly equally effective outcomes.

The ability of amino acid tracers to cross the BBB is a crucial advantage that transcends the limitations of contrast-enhanced MRI, which relies on leaky vascularity or compromise of the barrier for delivery of contrast enhancing agents. Compromise of the BBB by tumor growth does not appear to be a requirement for amino acid tracer permeability [31,32]. The intracellular uptake of amino acid tracers by the tumor cells is an active process facilitated by the L (large) transport system with subtypes LAT1 and LAT2 and is, therefore, more specific in reporting on live proliferating cells rather than structural changes [33–36]. This is further evidenced by dexamethasone or antiangiogenic treatment with bevacizumab not affecting amino acid uptake by brain tumors [37,38]. Prior or ongoing treatment with temozolomide, however, may impact the tumor-to-background ratio of amino acid uptake and must be considered in therapy response assessments [39].

The PET radiotracer [^{11}C-methyl]-L-methionine (^{11}C-MET) is the most well-established and utilized amino acid probe. ^{11}C-MET PET has been shown to be useful in delineating ependymomas, medulloblastoma, and astrocytomas in pediatric patients and can also effectively differentiate between radiation-induced brain tissue injury and tumor recurrence [40–43]. It is superior to MRI in differentiating tumor necrosis following gamma knife radiosurgery from recurrent tumor [44] and offers accurate tumor size correlation of WHO grades II and III meningiomas [45].

The short 20-min half-life of the ^{11}C, however, limits the practical implementation of ^{11}C-MET PET to highly specialized imaging centers with onsite cyclotron facilities. Consequently, ^{18}F-labeled amino acid tracers (with a half-life of 110 min), such as O-(2-^{18}F-fluoroethyl)-L-tyrosine (^{18}F-FET), 3,4-dihydroxy-6-^{18}F-fluoro-L-phenylalanine (^{18}F-FDOPA), and anti-1-amino-3-^{18}F-fluorocyclobutane-1-carboxylic acid (^{18}F-FACBC, fluciclovine, a non-natural amino acid) have been developed among many others to overcome this logistical limitation. Amino acid radiotracers, such as ^{18}F-FET, also exhibit tumor type-specific kinetics that can facilitate differential diagnosis of CNS tumors [46–51] (Figure 2). Metabolic changes in response to bevacizumab treatment of glioblastoma identified by ^{18}F-FET PET imaging occur earlier than morphologic changes providing a more immediate indication of tumor progression than changes detected by MRI [52]. Both ^{11}C-MET and ^{18}F-FACBC were able to differentiate tumor from equivocal lesions (edema) [53].

Although ^{18}F-FDOPA exhibits differential tumor uptake, its similarity to dopamine results in preferential accumulation in the striatum presenting an obstacle to accurately demarcating gliomas involving this region of the brain [54]. In comparison to ^{11}C-MET, ^{18}F-FACBC yields lower background accumulation levels (i.e., higher tumor-to-brain signal ratios) and, hence, higher detection sensitivity and specificity. ^{18}F-FDOPA PET metabolic tumor volume measurements at two weeks following antiangiogenic treatment are highly predictive of outcome. A first follow-up scan at two weeks predicted increased overall survival and progression-free survival. Responders identified based on ^{18}F-FDOPA-PET survived 3.5 times longer in contrast to responders identified by MRI who lived 1.5 times longer. ^{18}F-FDOPA PET is more accurate at identifying responders only two weeks following antiangiogenic treatment [55].

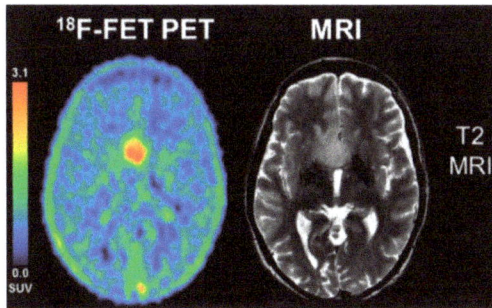

Figure 2. ^{18}F-FET PET summation image from 5–15 min is contrasted to a T2 MR image in this patient with a WHO III astrocytoma, IDH-wild-type, without 1p/19q co-deletion. This was a non-enhancing tumor on MRI. Note excellent tumor to background contrast for the FET PET image, with standard uptake values (SUV) of up to 3. Adapted from Unterrainer et al. *Eur. J. Nucl. Med. Mol. Imaging* 2018, 45, 1242 with permission [51].

4.2. Response to Therapy

Specific detection of early responses to therapy is a significant advantage of amino acid PET as it can specifically differentiate the cellular component of a tumor mass from inflammatory and necrotic lesions. Amino acid PET detects metabolic changes, which occur earlier than morphological changes detected by MRI in response to chemotherapy, radiation treatment, or antiangiogenic treatment. Favorable outcomes following three cycles of temozolomide treatment can be predicted by reduced ^{11}C-MET uptake in high-grade glioma patients [56,57]. Furthermore, the tumor-associated breakdown of the BBB does not appear to be a requirement for amino acid tracer permeability resulting in more accurate tumor boundary delineation even after vascular normalization by bevacizumab [31,32]. ^{18}F-FET and ^{18}F-FDOPA can be used to detect response to bevacizumab treatment before the appearance of morphological changes and thus provide a much earlier prediction of response [52,55,58]. In preclinical glioma models, ^{18}F-FET PET reports on early response to combination therapy with temozolomide, interferon-beta, or bevacizumab [59]. Management of recurrent high-grade glioma patients treated with bevacizumab and radiotherapy using ^{18}F-FET PET and MRI is cost effective and may potentially enhance the treatment quality. These outcomes are realized by the avoidance of unnecessary costs associated with overtreatment and unnecessary side effects [60].

4.3. Grade Differentiation

Differentiation between low and high-grade gliomas by imaging is a potentially beneficial diagnostic tool that would enhance the management of CNS tumors. Generally, WHO grade I and II tumors and a significant number of WHO grade III gliomas are contrast non-enhancing on MRI limiting tumor burden and therapy response assessments. Furthermore, differentiation between LGG and HGG by MRI remains challenging.

The utility of amino acid PET to differentiate between LGG and HGG based on the level of radiotracer uptake has been assessed in a limited number of studies. Generally, amino acid uptake is higher in grade III/IV gliomas compared to grade I/II [61–63]. [N-methyl-^{11}C] alpha-Methylaminoisobutyric acid

(^{11}C-MeAIB), an amino acid analog, can differentiate between low-grade and high-grade astrocytoma as determined by the tumor to normal brain uptake ratio [64]. Differential uptake of alpha-[^{11}C] methyl-L-tryptophan (^{11}C-AMT) can be used to identify LGGs, even if they are not contrast-enhancing on MRI, and distinguish them from HGGs based on tumor-to-cortex radiotracer uptake [65].

Despite differential LGG/HGG imaging in many studies, a recent systematic review of amino acid PET of LGGs has revealed a difficulty in interpretation due to inconsistencies in radiotracer uptake and different correlations between uptake ratios and LGG molecular status [66]. Radiotracer uptake intensities for each group can vary widely resulting in overlap and unreliable preoperative grading [67].

Kinetic PET imaging, however, may offer a better differentiating parameter based on the differing rate of radiotracer uptake by LGG and HGG. Time-activity curves of grade II tumors exhibit a slow, steady increase compared to the rapid uptake of amino acid radiotracers by grade III/IV tumors. Thus, LGG and HGG can be more reliably differentiated based on a dynamic or dual-time-point PET than static or endpoint imaging [47,48,61,68–70].

4.4. Prognostication

As an extension to its capacity to report on the early response to therapy, amino acid PET has proven to be a reliable prognostic tool in predicting overall and progression-free survival. Decreased ^{18}F-FET uptake following radiochemotherapy was associated with overall survival compared to increased or stable uptake [71,72]. True responders to bevacizumab treatment following recurrent glioma identified by ^{11}C-MET PET at eight weeks were predicted to have more favorable prognoses [73]. Determination of metabolic tumor volume of ^{11}C-MET-PET is prognostic of progression-free survival in high-grade glioma patients [74] and predictive of patients developing recurrent malignant glioma [75].

Amino acid uptake may also serve as a reliable predictor of survival in patients with LGGs. ^{18}F-FET PET combined with anatomic MRI predicted outcome and progression-free survival [76]. Kinetics of amino acid uptake have also been demonstrated to be useful in predicting regions of malignant transformation and progression within LGG [70,77]. Detection of metabolic abnormalities in LGGs by amino acid PET following treatment may also serve as a predictor of tumor recurrence [78].

4.5. Biopsy Guidance

Diagnostic inaccuracies based on image interpretation carry risk and compromise CNS tumor therapy outcomes. Biopsy provides diagnostic accuracy at a higher rate than neuroimaging and is essential to tumor grading and treatment planning. Therefore, image-guided stereotactic biopsy of tumor volumes is crucial to achieving higher diagnostic yields. Delineation of tumor extent and identification of regions of higher grade and density with ^{18}F-FET and ^{18}F-DOPA PET proved useful for biopsy guidance and resection planning in most cerebral glioma cases compared to ^{18}F-FDG [67,79–81]. Needle guidance to hypermetabolic foci by PET produce a higher diagnostic yield than MRI reducing sampling to a single trajectory. Reducing sampling frequency through accurate image data is particularly important to minimize the risks associated with a biopsy of intrinsic infiltrative brainstem malignancies [82].

4.6. Case Studies Validating Amino Acid PET

Some case studies illustrate the advantages of amino acid PET over MRI in delineating tumor extent, informing on response to therapy, and detecting disease recurrence. Amino acid uptake has also been shown to be consistent with exhibited symptoms and biopsy results. A unique case study involving a 37-year old woman with a "butterfly" glioblastoma exemplifies tumor extent delineation by amino acid PET. MRI findings were equivocal and imprecise, indicating bifocal growth with slight contrast enhancement despite the large extent of the bilateral tumor mass. ^{11}C-MET PET was more diagnostically precise than contrast-enhanced MRI in delineating a broad and continuous tumor mass [83].

Amino acid PET has also proved valuable in detecting recurrence and differentiating it from pseudoprogression as confirmed by biopsy. Detection of recurrence by PET was illustrated by high ^{18}F-DOPA uptake in the left parietal lobe of a 46-year-old patient, where resection of a glioma had taken place two years earlier. ^{18}F-FDG uptake was not elevated and did not indicate any abnormalities. Recurrence of the tumor in this patient was confirmed by biopsy demonstrating the accuracy of amino acid PET in detecting metabolic tumor volume [84]. Similarly, pseudoprogression has been reliably determined by amino acid PET in other case studies. Morphological changes on MRI suggested recurrent tumor eight months following radiochemotherapy in a glioblastoma patient. ^{18}F-FET PET, however, was negative for focal uptake and, indeed, histopathology of the resected tumor revealed necrotic tissue consistent with pseudoprogression. The lesion regressed in follow-up MRI further confirming the diagnosis [85].

In contrast to MRI, consistency of amino acid uptake with patient symptoms has been demonstrated in a number of case studies. Vision problems exhibited by 16-year old female optic pathway glioma patient correlated with an increase in ^{11}C-AMT uptake indicating recurrence. ^{11}C-AMT uptake decreased upon chemotherapy and radiotherapy and correlated with a symptomatic improvement [86]. Likewise, ^{18}F-MET PET was able to detect recurrence of a benign oligodendroglioma in a patient exhibiting recurrent temporal epileptic seizures 15 months following resection. ^{18}F-MET PET uptake was high in the region of the previous tumor. Secondary resection and biopsy confirmed recurrence as predicted by PET [87]. In both of these cases, MRI did not reveal morphological changes indicating tumor recurrence and was inconsistent with exhibited patient symptoms.

5. Combined PET/MRI for the Management of CNS Tumors

The strength of MRI in providing high anatomical resolution of soft brain tissue coupled with the specificity of amino acid PET in delineating CNS tumors promote combining the modalities to improve diagnostic specificity in the management of CNS malignancies. When combined with MRI, ^{11}C-MET-PET has been shown to be a reliable identifier of true responders to bevacizumab therapy with favorable prognoses [73,88,89]. Furthermore, improved diagnostic confidence of primary low-grade astrocytoma and high-grade astrocytomas [90] and reliable early differentiation between tumor recurrence and radionecrosis in high-grade glioma patients have been demonstrated [91]. The benefits of combined PET/MRI have also been extended to pediatric patient management and found to be well-tolerated and have been recommended for early prediction of response [92].

6. Conclusions and Future Perspectives

Metabolic changes detected by amino acid PET occur in response to therapy sooner than morphological and structural changes providing an earlier response to therapy assessment and prognosis. Moreover, metabolic uptake of radiolabeled amino acids is specific to the proliferating cellular component of the tumor, occurs independently of the status of the blood-brain barrier, and excludes non-cellular lesions created by necrosis, edema, and inflammation. As summarized in Figure 3, these advantages offered by amino acid PET translate into precise tumor boundary delineation and early treatment response assessments that transcend MRI-based limitations of pseudoprogression and pseudoresponse.

The advantages provided by amino acid PET justify its future implementation as a standard procedure in defining biopsy sites, planning surgical resection and radiotherapy, prognostication, and treatment response assessment [93,94]. These tangible benefits have been realized leading to the recommendation by the Response Assessment Neuro-Oncology (RANO) working group for the inclusion of amino acid PET in the management and diagnosis of brain tumors in conjunction with conventional MRI [95]. Complementing the limitations of MRI with the strengths of amino acid PET will lead to significant improvements in all aspects of CNS tumor management.

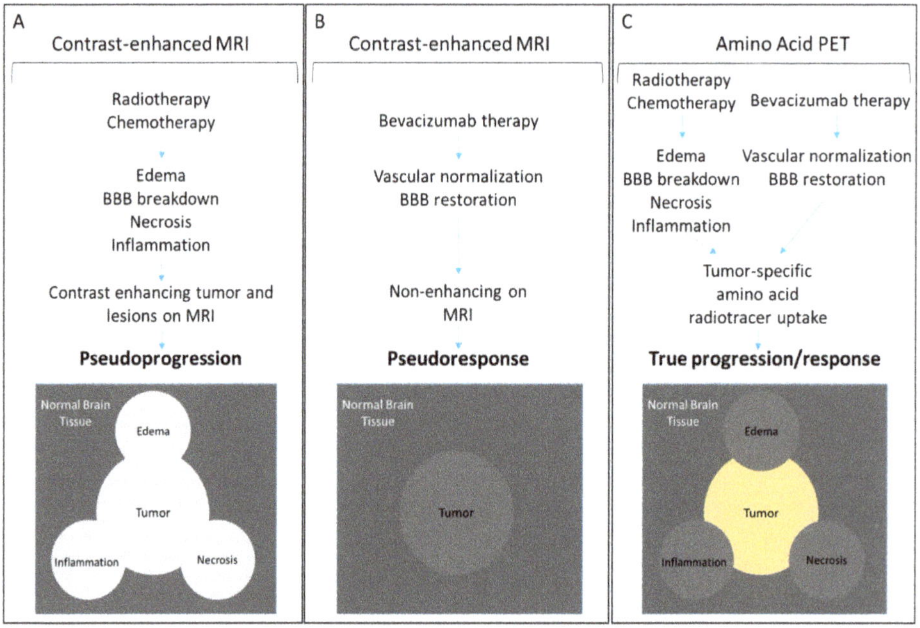

Figure 3. Schematic representation of MRI and PET outcomes following treatment of CNS tumors. (**A**) Radiotherapy and chemotherapy of CNS tumors can lead to edema, necrosis, inflammation, and breakdown of the BBB creating contrast-enhancing lesions that obscure tumor boundaries and lead to pseudoprogression. (**B**) Conversely, vascular normalization following bevacizumab treatment, without an associated anti-tumor response, can diminish contrast enhancement yielding a false appearance of treatment response or pseudoresponse. (**C**) Amino acid PET reports specifically on the accumulation of the radiotracers within the cellular component of the tumor mass reflecting correct tumor boundaries and immediate metabolic changes in response to therapy.

Author Contributions: Conceptualization: A.M.N., J.M.J., and D.S.; writing—original draft preparation: A.M.N.; writing—review, and editing: A.M.N., J.M.J., and D.S.

Funding: This research received no external funding.

Conflicts of Interest: The authors declare no conflict of interest.

References

1. Cancer Facts & Figures 2018. Available online: https://www.cancer.org/research/cancer-facts-statistics/all-cancer-facts-figures/cancer-facts-figures-2018.html (accessed on 10 September 2018).
2. Ostrom, Q.T.; Gittleman, H.; Liao, P.; Vecchione-Koval, T.; Wolinsky, Y.; Kruchko, C.; Barnholtz-Sloan, J.S. CBTRUS Statistical Report: Primary brain and other central nervous system tumors diagnosed in the United States in 2010–2014. *Neuro-Oncology* **2017**, *19* (Suppl. S5), v1–v88. [CrossRef] [PubMed]
3. Nayak, L.; Lee, E.Q.; Wen, P.Y. Epidemiology of brain metastases. *Curr. Oncol. Rep.* **2012**, *14*, 48–54. [CrossRef] [PubMed]
4. Louis, D.N.; Perry, A.; Reifenberger, G.; von Deimling, A.; Figarella-Branger, D.; Cavenee, W.K.; Ohgaki, H.; Wiestler, O.D.; Kleihues, P.; Ellison, D.W. The 2016 World Health Organization Classification of Tumors of the Central Nervous System: A summary. *Acta Neuropathol.* **2016**, *131*, 803–820. [CrossRef] [PubMed]
5. Ohgaki, H.; Kleihues, P. Population-based studies on incidence, survival rates, and genetic alterations in astrocytic and oligodendroglial gliomas. *J. Neuropathol. Exp. Neurol.* **2005**, *64*, 479–489. [CrossRef] [PubMed]

6.   Abbott, N.J. Blood-brain barrier structure and function and the challenges for CNS drug delivery. *J. Inherit. Metab. Dis.* **2013**, *36*, 437–449. [CrossRef] [PubMed]

7.   Da Fonseca, A.C.; Matias, D.; Garcia, C.; Amaral, R.; Geraldo, L.H.; Freitas, C.; Lima, F.R. The impact of microglial activation on blood-brain barrier in brain diseases. *Front. Cell. Neurosci.* **2014**, *8*, 362. [CrossRef] [PubMed]

8.   Abbott, N.J.; Ronnback, L.; Hansson, E. Astrocyte-endothelial interactions at the blood-brain barrier. *Nat. Rev. Neurosci.* **2006**, *7*, 41–53. [CrossRef] [PubMed]

9.   Davies, D.C. Blood-brain barrier breakdown in septic encephalopathy and brain tumours. *J. Anat.* **2002**, *200*, 639–646. [CrossRef] [PubMed]

10. Papadopoulos, M.C.; Saadoun, S.; Davies, D.C.; Bell, B.A. Emerging molecular mechanisms of brain tumour oedema. *Br. J. Neurosurg.* **2001**, *15*, 101–108. [CrossRef] [PubMed]

11. Margolin, K.; Ernstoff, M.S.; Hamid, O.; Lawrence, D.; McDermott, D.; Puzanov, I.; Wolchok, J.D.; Clark, J.I.; Sznol, M.; Logan, T.F.; et al. Ipilimumab in patients with melanoma and brain metastases: An open-label, phase 2 trial. *Lancet. Oncol.* **2012**, *13*, 459–465. [CrossRef]

12. Jain, R.; Griffith, B.; Alotaibi, F.; Zagzag, D.; Fine, H.; Golfinos, J.; Schultz, L. Glioma Angiogenesis and Perfusion Imaging: Understanding the Relationship between Tumor Blood Volume and Leakiness with Increasing Glioma Grade. *AJNR Am. J. Neuroradiol.* **2015**, *36*, 2030–2035. [CrossRef] [PubMed]

13. Jain, R.K.; di Tomaso, E.; Duda, D.G.; Loeffler, J.S.; Sorensen, A.G.; Batchelor, T.T. Angiogenesis in brain tumours. *Nat. Rev. Neurosci.* **2007**, *8*, 610–622. [CrossRef] [PubMed]

14. Lyle, L.T.; Lockman, P.R.; Adkins, C.E.; Mohammad, A.S.; Sechrest, E.; Hua, E.; Palmieri, D.; Liewehr, D.J.; Steinberg, S.M.; Kloc, W.; et al. Alterations in Pericyte Subpopulations Are Associated with Elevated Blood-Tumor Barrier Permeability in Experimental Brain Metastasis of Breast Cancer. *Clin. Cancer Res. Off. J. Am. Assoc. Cancer Res.* **2016**, *22*, 5287–5299. [CrossRef] [PubMed]

15. Russell, S.M.; Elliott, R.; Forshaw, D.; Golfinos, J.G.; Nelson, P.K.; Kelly, P.J. Glioma vascularity correlates with reduced patient survival and increased malignancy. *Surg. Neurol.* **2009**, *72*, 242–246. [CrossRef] [PubMed]

16. Leon, S.P.; Folkerth, R.D.; Black, P.M. Microvessel density is a prognostic indicator for patients with astroglial brain tumors. *Cancer* **1996**, *77*, 362–372. [CrossRef]

17. Wesseling, P.; van der Laak, J.A.; Link, M.; Teepen, H.L.; Ruiter, D.J. Quantitative analysis of microvascular changes in diffuse astrocytic neoplasms with increasing grade of malignancy. *Hum. Pathol.* **1998**, *29*, 352–358. [CrossRef]

18. Chinot, O.L.; Wick, W.; Mason, W.; Henriksson, R.; Saran, F.; Nishikawa, R.; Carpentier, A.F.; Hoang-Xuan, K.; Kavan, P.; Cernea, D.; et al. Bevacizumab plus radiotherapy-temozolomide for newly diagnosed glioblastoma. *N. Engl. J. Med.* **2014**, *370*, 709–722. [CrossRef] [PubMed]

19. Lamborn, K.R.; Yung, W.K.; Chang, S.M.; Wen, P.Y.; Cloughesy, T.F.; DeAngelis, L.M.; Robins, H.I.; Lieberman, F.S.; Fine, H.A.; Fink, K.L.; et al. Progression-free survival: An important end point in evaluating therapy for recurrent high-grade gliomas. *Neuro-Oncology* **2008**, *10*, 162–170. [CrossRef] [PubMed]

20. Macdonald, D.R.; Cascino, T.L.; Schold, S.C., Jr.; Cairncross, J.G. Response criteria for phase II studies of supratentorial malignant glioma. *J. Clin. Oncol. Off. J. Am. Soc. Clin. Oncol.* **1990**, *8*, 1277–1280. [CrossRef] [PubMed]

21. Chinot, O.L.; Macdonald, D.R.; Abrey, L.E.; Zahlmann, G.; Kerloeguen, Y.; Cloughesy, T.F. Response assessment criteria for glioblastoma: Practical adaptation and implementation in clinical trials of antiangiogenic therapy. *Curr. Neurol. Neurosci. Rep.* **2013**, *13*, 347. [CrossRef] [PubMed]

22. Chiou, V.L.; Burotto, M. Pseudoprogression and Immune-Related Response in Solid Tumors. *J. Clin. Oncol. Off. J. Am. Soc. Clin. Oncol.* **2015**, *33*, 3541–3543. [CrossRef] [PubMed]

23. Jain, R.K. Antiangiogenic therapy for cancer: Current and emerging concepts. *Oncology* **2005**, *19*, 7–16. [PubMed]

24. Hygino da Cruz, L.C., Jr.; Rodriguez, I.; Domingues, R.C.; Gasparetto, E.L.; Sorensen, A.G. Pseudoprogression and pseudoresponse: Imaging challenges in the assessment of posttreatment glioma. *AJNR Am. J. Neuroradiol.* **2011**, *32*, 1978–1985. [CrossRef] [PubMed]

25. Herbel, C.; Patsoukis, N.; Bardhan, K.; Seth, P.; Weaver, J.D.; Boussiotis, V.A. Clinical significance of T cell metabolic reprogramming in cancer. *Clin. Transl. Med.* **2016**, *5*, 29. [CrossRef] [PubMed]

26. Bollineni, V.R.; Kramer, G.M.; Jansma, E.P.; Liu, Y.; Oyen, W.J. A systematic review on [(18)F]FLT-PET uptake as a measure of treatment response in cancer patients. *Eur. J. Cancer* **2016**, *55*, 81–97. [CrossRef] [PubMed]

27. Chen, W.; Delaloye, S.; Silverman, D.H.; Geist, C.; Czernin, J.; Sayre, J.; Satyamurthy, N.; Pope, W.; Lai, A.; Phelps, M.E.; et al. Predicting treatment response of malignant gliomas to bevacizumab and irinotecan by imaging proliferation with [18F] fluorothymidine positron emission tomography: A pilot study. *J. Clin. Oncol. Off. J. Am. Soc. Clin. Oncol.* **2007**, *25*, 4714–4721. [CrossRef] [PubMed]

28. Schwarzenberg, J.; Czernin, J.; Cloughesy, T.F.; Ellingson, B.M.; Pope, W.B.; Geist, C.; Dahlbom, M.; Silverman, D.H.; Satyamurthy, N.; Phelps, M.E.; et al. 3′-deoxy-3′-18F-fluorothymidine PET and MRI for early survival predictions in patients with recurrent malignant glioma treated with bevacizumab. *J. Nucl. Med. Off. Publ. Soc. Nucl. Med.* **2012**, *53*, 29–36. [CrossRef] [PubMed]

29. Wardak, M.; Schiepers, C.; Cloughesy, T.F.; Dahlbom, M.; Phelps, M.E.; Huang, S.C. (1)(8)F-FLT and (1)(8)F-FDOPA PET kinetics in recurrent brain tumors. *Eur. J. Nucl. Med. Mol. Imaging* **2014**, *41*, 1199–1209. [CrossRef] [PubMed]

30. Nowosielski, M.; DiFranco, M.D.; Putzer, D.; Seiz, M.; Recheis, W.; Jacobs, A.H.; Stockhammer, G.; Hutterer, M. An intra-individual comparison of MRI, [18F]-FET and [18F]-FLT PET in patients with high-grade gliomas. *PLoS ONE* **2014**, *9*, e95830. [CrossRef] [PubMed]

31. Rapp, M.; Floeth, F.W.; Felsberg, J.; Steiger, H.J.; Sabel, M.; Langen, K.J.; Galldiks, N. Clinical value of O-(2-[(18)F]-fluoroethyl)-L-tyrosine positron emission tomography in patients with low-grade glioma. *Neurosurg. Focus* **2013**, *34*, E3. [CrossRef] [PubMed]

32. Unterrainer, M.; Schweisthal, F.; Suchorska, B.; Wenter, V.; Schmid-Tannwald, C.; Fendler, W.P.; Schuller, U.; Bartenstein, P.; Tonn, J.C.; Albert, N.L. Serial 18F-FET PET Imaging of Primarily 18F-FET-Negative Glioma: Does It Make Sense? *J. Nucl. Med. Off. Publ. Soc. Nucl. Med.* **2016**, *57*, 1177–1182. [CrossRef] [PubMed]

33. Okubo, S.; Zhen, H.N.; Kawai, N.; Nishiyama, Y.; Haba, R.; Tamiya, T. Correlation of L-methyl-11C-methionine (MET) uptake with L-type amino acid transporter 1 in human gliomas. *J. Neuro-Oncol.* **2010**, *99*, 217–225. [CrossRef] [PubMed]

34. Youland, R.S.; Kitange, G.J.; Peterson, T.E.; Pafundi, D.H.; Ramiscal, J.A.; Pokorny, J.L.; Giannini, C.; Laack, N.N.; Parney, I.F.; Lowe, V.J.; et al. The role of LAT1 in (18)F-DOPA uptake in malignant gliomas. *J. Neuro-Oncol.* **2013**, *111*, 11–18. [CrossRef] [PubMed]

35. Habermeier, A.; Graf, J.; Sandhofer, B.F.; Boissel, J.P.; Roesch, F.; Closs, E.I. System L amino acid transporter LAT1 accumulates O-(2-fluoroethyl)-L-tyrosine (FET). *Amino Acids* **2015**, *47*, 335–344. [CrossRef] [PubMed]

36. Dadone-Montaudie, B.; Ambrosetti, D.; Dufour, M.; Darcourt, J.; Almairac, F.; Coyne, J.; Virolle, T.; Humbert, O.; Burel-Vandenbos, F. [18F] FDOPA standardized uptake values of brain tumors are not exclusively dependent on LAT1 expression. *PLoS ONE* **2017**, *12*, e0184625. [CrossRef] [PubMed]

37. Stegmayr, C.; Bandelow, U.; Oliveira, D.; Lohmann, P.; Willuweit, A.; Filss, C.; Galldiks, N.; Lubke, J.H.; Shah, N.J.; Ermert, J.; et al. Influence of blood-brain barrier permeability on O-(2-(18)F-fluoroethyl)-L-tyrosine uptake in rat gliomas. *Eur. J. Nucl. Med. Mol. Imaging* **2017**, *44*, 408–416. [CrossRef] [PubMed]

38. Stegmayr, C.; Oliveira, D.; Niemietz, N.; Willuweit, A.; Lohmann, P.; Galldiks, N.; Shah, N.J.; Ermert, J.; Langen, K.J. Influence of Bevacizumab on Blood-Brain Barrier Permeability and O-(2-(18)F-Fluoroethyl)-L-Tyrosine Uptake in Rat Gliomas. *J. Nucl. Med. Off. Publ. Soc. Nucl. Med.* **2017**, *58*, 700–705.

39. Carideo, L.; Minniti, G.; Mamede, M.; Scaringi, C.; Russo, I.; Scopinaro, F.; Cicone, F. (18)F-DOPA uptake parameters in glioma: Effects of patients' characteristics and prior treatment history. *Br. J. Radiol.* **2018**, *91*, 20170847. [CrossRef] [PubMed]

40. O'Tuama, L.A.; Phillips, P.C.; Strauss, L.C.; Carson, B.C.; Uno, Y.; Smith, Q.R.; Dannals, R.F.; Wilson, A.A.; Ravert, H.T.; Loats, S.; et al. Two-phase [11C]L-methionine PET in childhood brain tumors. *Pediatric Neurol.* **1990**, *6*, 163–170. [CrossRef]

41. Kits, A.; Martin, H.; Sanchez-Crespo, A.; Delgado, A.F. Diagnostic accuracy of (11)C-methionine PET in detecting neuropathologically confirmed recurrent brain tumor after radiation therapy. *Ann. Nucl. Med.* **2018**, *32*, 132–141. [CrossRef] [PubMed]

42. Xu, W.; Gao, L.; Shao, A.; Zheng, J.; Zhang, J. The performance of 11C-Methionine PET in the differential diagnosis of glioma recurrence. *Oncotarget* **2017**, *8*, 91030–91039. [CrossRef] [PubMed]

43. Yomo, S.; Oguchi, K. Prospective study of (11)C-methionine PET for distinguishing between recurrent brain metastases and radiation necrosis: Limitations of diagnostic accuracy and long-term results of salvage treatment. *BMC Cancer* **2017**, *17*, 713. [CrossRef] [PubMed]

44. Tomura, N.; Kokubun, M.; Saginoya, T.; Mizuno, Y.; Kikuchi, Y. Differentiation between Treatment-Induced Necrosis and Recurrent Tumors in Patients with Metastatic Brain Tumors: Comparison among (11)C-Methionine-PET, FDG-PET, MR Permeability Imaging, and MRI-ADC-Preliminary Results. *AJNR Am. J. Neuroradiol.* **2017**, *38*, 1520–1527. [CrossRef] [PubMed]

45. Tomura, N.; Saginoya, T.; Goto, H. 11C-Methionine Positron Emission Tomography/Computed Tomography Versus 18F-Fluorodeoxyglucose Positron Emission Tomography/Computed Tomography in Evaluation of Residual or Recurrent World Health Organization Grades II and III Meningioma After Treatment. *J. Comput. Assist. Tomogr.* **2018**, *42*, 517–521. [CrossRef] [PubMed]

46. Kratochwil, C.; Combs, S.E.; Leotta, K.; Afshar-Oromieh, A.; Rieken, S.; Debus, J.; Haberkorn, U.; Giesel, F.L. Intra-individual comparison of (1)(8)F-FET and (1)(8)F-DOPA in PET imaging of recurrent brain tumors. *Neuro-Oncology* **2014**, *16*, 434–440. [CrossRef] [PubMed]

47. Calcagni, M.L.; Galli, G.; Giordano, A.; Taralli, S.; Anile, C.; Niesen, A.; Baum, R.P. Dynamic O-(2-[18F]fluoroethyl)-L-tyrosine (F-18 FET) PET for glioma grading: Assessment of individual probability of malignancy. *Clin. Nucl. Med.* **2011**, *36*, 841–847. [CrossRef] [PubMed]

48. Popperl, G.; Kreth, F.W.; Mehrkens, J.H.; Herms, J.; Seelos, K.; Koch, W.; Gildehaus, F.J.; Kretzschmar, H.A.; Tonn, J.C.; Tatsch, K. FET PET for the evaluation of untreated gliomas: Correlation of FET uptake and uptake kinetics with tumour grading. *Eur. J. Nucl. Med. Mol. Imaging* **2007**, *34*, 1933–1942. [CrossRef] [PubMed]

49. Weckesser, M.; Langen, K.J.; Rickert, C.H.; Kloska, S.; Straeter, R.; Hamacher, K.; Kurlemann, G.; Wassmann, H.; Coenen, H.H.; Schober, O. O-(2-[18F]fluoroethyl)-L-tyrosine PET in the clinical evaluation of primary brain tumours. *Eur. J. Nucl. Med. Mol. Imaging* **2005**, *32*, 422–429. [CrossRef] [PubMed]

50. Moulin-Romsee, G.; D'Hondt, E.; de Groot, T.; Goffin, J.; Sciot, R.; Mortelmans, L.; Menten, J.; Bormans, G.; Van Laere, K. Non-invasive grading of brain tumours using dynamic amino acid PET imaging: Does it work for 11C-methionine? *Eur. J. Nucl. Med. Mol. Imaging* **2007**, *34*, 2082–2087. [CrossRef] [PubMed]

51. Unterrainer, M.; Winkelmann, I.; Suchorska, B.; Giese, A.; Wenter, V.; Kreth, F.W.; Herms, J.; Bartenstein, P.; Tonn, J.C.; Albert, N.L. Correction to: Biological tumour volumes of gliomas in early and standard 20-40 min (18)F-FET PET images differ according to IDH mutation status. *Eur. J. Nucl. Med. Mol. Imaging* **2018**, *45*, 1078. [CrossRef] [PubMed]

52. Galldiks, N.; Rapp, M.; Stoffels, G.; Dunkl, V.; Sabel, M.; Langen, K.J. Earlier diagnosis of progressive disease during bevacizumab treatment using O-(2-18F-fluoroethyl)-L-tyrosine positron emission tomography in comparison with magnetic resonance imaging. *Mol. Imaging* **2013**, *12*, 273–276. [CrossRef] [PubMed]

53. Tsuyuguchi, N.; Terakawa, Y.; Uda, T.; Nakajo, K.; Kanemura, Y. Diagnosis of Brain Tumors Using Amino Acid Transport PET Imaging with (18)F-fluciclovine: A Comparative Study with L-methyl-(11)C-methionine PET Imaging. *Asia Ocean. J. Nucl. Med. Boil.* **2017**, *5*, 85–94.

54. Cicone, F.; Filss, C.P.; Minniti, G.; Rossi-Espagnet, C.; Papa, A.; Scaringi, C.; Galldiks, N.; Bozzao, A.; Shah, N.J.; Scopinaro, F.; et al. Volumetric assessment of recurrent or progressive gliomas: Comparison between F-DOPA PET and perfusion-weighted MRI. *Eur. J. Nucl. Med. Mol. Imaging* **2015**, *42*, 905–915. [CrossRef] [PubMed]

55. Schwarzenberg, J.; Czernin, J.; Cloughesy, T.F.; Ellingson, B.M.; Pope, W.B.; Grogan, T.; Elashoff, D.; Geist, C.; Silverman, D.H.; Phelps, M.E.; et al. Treatment response evaluation using 18F-FDOPA PET in patients with recurrent malignant glioma on bevacizumab therapy. *Clin. Cancer Res. Off. J. Am. Assoc. Cancer Res.* **2014**, *20*, 3550–3559. [CrossRef] [PubMed]

56. Galldiks, N.; Kracht, L.W.; Burghaus, L.; Thomas, A.; Jacobs, A.H.; Heiss, W.D.; Herholz, K. Use of 11C-methionine PET to monitor the effects of temozolomide chemotherapy in malignant gliomas. *Eur. J. Nucl. Med. Mol. Imaging* **2006**, *33*, 516–524. [CrossRef] [PubMed]

57. Galldiks, N.; Kracht, L.W.; Burghaus, L.; Ullrich, R.T.; Backes, H.; Brunn, A.; Heiss, W.D.; Jacobs, A.H. Patient-tailored, imaging-guided, long-term temozolomide chemotherapy in patients with glioblastoma. *Mol. Imaging* **2010**, *9*, 40–46. [CrossRef] [PubMed]

58. Galldiks, N.; Dunkl, V.; Ceccon, G.; Tscherpel, C.; Stoffels, G.; Law, I.; Henriksen, O.M.; Muhic, A.; Poulsen, H.S.; Steger, J.; et al. Early treatment response evaluation using FET PET compared to MRI in glioblastoma patients at first progression treated with bevacizumab plus lomustine. *Eur. J. Nucl. Med. Mol. Imaging* **2018**, *45*, 2377–2386. [CrossRef] [PubMed]

59. Ono, T.; Sasajima, T.; Doi, Y.; Oka, S.; Ono, M.; Kanagawa, M.; Baden, A.; Mizoi, K.; Shimizu, H. Amino acid PET tracers are reliable markers of treatment responses to single-agent or combination therapies including temozolomide, interferon-beta, and/or bevacizumab for glioblastoma. *Nucl. Med. Boil.* **2015**, *42*, 598–607. [CrossRef] [PubMed]

60. Heinzel, A.; Muller, D.; Langen, K.J.; Blaum, M.; Verburg, F.A.; Mottaghy, F.M.; Galldiks, N. The use of O-(2-18F-fluoroethyl)-L-tyrosine PET for treatment management of bevacizumab and irinotecan in patients with recurrent high-grade glioma: a cost-effectiveness analysis. *J. Nucl. Med. Off. Publ. Soc. Nucl. Med.* **2013**, *54*, 1217–1222. [CrossRef] [PubMed]

61. Jansen, N.L.; Graute, V.; Armbruster, L.; Suchorska, B.; Lutz, J.; Eigenbrod, S.; Cumming, P.; Bartenstein, P.; Tonn, J.C.; Kreth, F.W.; et al. MRI-suspected low-grade glioma: Is there a need to perform dynamic FET PET? *Eur. J. Nucl. Med. Mol. Imaging* **2012**, *39*, 1021–1029. [CrossRef] [PubMed]

62. Lopez, W.O.; Cordeiro, J.G.; Albicker, U.; Doostkam, S.; Nikkhah, G.; Kirch, R.D.; Trippel, M.; Reithmeier, T. Correlation of (18)F-fluoroethyl tyrosine positron-emission tomography uptake values and histomorphological findings by stereotactic serial biopsy in newly diagnosed brain tumors using a refined software tool. *OncoTargets Ther.* **2015**, *8*, 3803–3815. [CrossRef] [PubMed]

63. Cicuendez, M.; Lorenzo-Bosquet, C.; Cuberas-Borros, G.; Martinez-Ricarte, F.; Cordero, E.; Martinez-Saez, E.; Castell-Conesa, J.; Sahuquillo, J. Role of [(11)C] methionine positron emission tomography in the diagnosis and prediction of survival in brain tumours. *Clin. Neurol. Neurosurg.* **2015**, *139*, 328–333. [CrossRef] [PubMed]

64. Nishii, R.; Higashi, T.; Kagawa, S.; Arimoto, M.; Kishibe, Y.; Takahashi, M.; Yamada, S.; Saiki, M.; Arakawa, Y.; Yamauchi, H.; et al. Differential Diagnosis between Low-Grade and High-Grade Astrocytoma Using System A Amino Acid Transport PET Imaging with C-11-MeAIB: A Comparison Study with C-11-Methionine PET Imaging. *Contrast Media Mol. Imaging* **2018**, *2018*, 1292746. [CrossRef] [PubMed]

65. Juhasz, C.; Muzik, O.; Chugani, D.C.; Chugani, H.T.; Sood, S.; Chakraborty, P.K.; Barger, G.R.; Mittal, S. Differential kinetics of alpha-[(1)(1)C]methyl-L-tryptophan on PET in low-grade brain tumors. *J. Neuro-Oncol.* **2011**, *102*, 409–415. [CrossRef] [PubMed]

66. Naslund, O.; Smits, A.; Forander, P.; Laesser, M.; Bartek, J., Jr.; Gempt, J.; Liljegren, A.; Daxberg, E.L.; Jakola, A.S. Amino acid tracers in PET imaging of diffuse low-grade gliomas: A systematic review of preoperative applications. *Acta Neurochir.* **2018**, *160*, 1451–1460. [CrossRef] [PubMed]

67. Pauleit, D.; Stoffels, G.; Bachofner, A.; Floeth, F.W.; Sabel, M.; Herzog, H.; Tellmann, L.; Jansen, P.; Reifenberger, G.; Hamacher, K.; et al. Comparison of (18)F-FET and (18)F-FDG PET in brain tumors. *Nucl. Med. Boil.* **2009**, *36*, 779–787. [CrossRef] [PubMed]

68. Lohmann, P.; Herzog, H.; Rota Kops, E.; Stoffels, G.; Judov, N.; Filss, C.; Galldiks, N.; Tellmann, L.; Weiss, C.; Sabel, M.; et al. dual-time-point O-(2-[(18)F]fluoroethyl)-L-tyrosine PET for grading of cerebral gliomas. *Eur. Radiol.* **2015**, *25*, 3017–3024. [CrossRef] [PubMed]

69. Albert, N.L.; Winkelmann, I.; Suchorska, B.; Wenter, V.; Schmid-Tannwald, C.; Mille, E.; Todica, A.; Brendel, M.; Tonn, J.C.; Bartenstein, P.; et al. Early static (18)F-FET-PET scans have a higher accuracy for glioma grading than the standard 20–40 min scans. *Eur. J. Nucl. Med. Mol. Imaging* **2016**, *43*, 1105–1114. [CrossRef] [PubMed]

70. Jansen, N.L.; Suchorska, B.; Wenter, V.; Eigenbrod, S.; Schmid-Tannwald, C.; Zwergal, A.; Niyazi, M.; Drexler, M.; Bartenstein, P.; Schnell, O.; et al. Dynamic 18F-FET PET in newly diagnosed astrocytic low-grade glioma identifies high-risk patients. *J. Nucl. Med. Off. Publ. Soc. Nucl. Med.* **2014**, *55*, 198–203. [CrossRef] [PubMed]

71. Piroth, M.D.; Holy, R.; Pinkawa, M.; Stoffels, G.; Kaiser, H.J.; Galldiks, N.; Herzog, H.; Coenen, H.H.; Eble, M.J.; Langen, K.J. Prognostic impact of postoperative, pre-irradiation (18)F-fluoroethyl-L-tyrosine uptake in glioblastoma patients treated with radiochemotherapy. *Radiother. Oncol. J. Eur. Soc. Ther. Radiol. Oncol.* **2011**, *99*, 218–224. [CrossRef] [PubMed]

72. Galldiks, N.; Langen, K.J.; Holy, R.; Pinkawa, M.; Stoffels, G.; Nolte, K.W.; Kaiser, H.J.; Filss, C.P.; Fink, G.R.; Coenen, H.H.; et al. Assessment of treatment response in patients with glioblastoma using O-(2-18F-fluoroethyl)-L-tyrosine PET in comparison to MRI. *J. Nucl. Med. Off. Publ. Soc. Nucl. Med.* **2012**, *53*, 1048–1057. [CrossRef] [PubMed]

73. Beppu, T.; Terasaki, K.; Sasaki, T.; Sato, Y.; Tomabechi, M.; Kato, K.; Sasaki, M.; Ogasawara, K. MRI and 11C-methyl-L-methionine PET Differentiate Bevacizumab True Responders After Initiating Therapy for Recurrent Glioblastoma. *Clin. Nucl. Med.* **2016**, *41*, 852–857. [CrossRef] [PubMed]
74. Yoo, M.Y.; Paeng, J.C.; Cheon, G.J.; Lee, D.S.; Chung, J.K.; Kim, E.E.; Kang, K.W. Prognostic Value of Metabolic Tumor Volume on (11)C-Methionine PET in Predicting Progression-Free Survival in High-Grade Glioma. *Nucl. Med. Mol. Imaging* **2015**, *49*, 291–297. [CrossRef] [PubMed]
75. Jung, T.Y.; Min, J.J.; Bom, H.S.; Jung, S.; Kim, I.Y.; Lim, S.H.; Kim, D.Y.; Kwon, S.Y. Prognostic value of post-treatment metabolic tumor volume from (11)C-methionine PET/CT in recurrent malignant glioma. *Neurosurg. Rev.* **2017**, *40*, 223–229. [CrossRef] [PubMed]
76. Floeth, F.W.; Sabel, M.; Stoffels, G.; Pauleit, D.; Hamacher, K.; Steiger, H.J.; Langen, K.J. Prognostic value of 18F-fluoroethyl-L-tyrosine PET and MRI in small nonspecific incidental brain lesions. *J. Nucl. Med. Off. Publ. Soc. Nucl. Med.* **2008**, *49*, 730–737. [CrossRef] [PubMed]
77. Galldiks, N.; Stoffels, G.; Ruge, M.I.; Rapp, M.; Sabel, M.; Reifenberger, G.; Erdem, Z.; Shah, N.J.; Fink, G.R.; Coenen, H.H.; et al. Role of O-(2-18F-fluoroethyl)-L-tyrosine PET as a diagnostic tool for detection of malignant progression in patients with low-grade glioma. *J. Nucl. Med. Off. Publ. Soc. Nucl. Med.* **2013**, *54*, 2046–2054. [CrossRef] [PubMed]
78. Morana, G.; Piccardo, A.; Garre, M.L.; Nozza, P.; Consales, A.; Rossi, A. Multimodal magnetic resonance imaging and 18F-L-dihydroxyphenylalanine positron emission tomography in early characterization of pseudoresponse and nonenhancing tumor progression in a pediatric patient with malignant transformation of ganglioglioma treated with bevacizumab. *J. Clin. Oncol. Off. J. Am. Soc. Clin. Oncol.* **2013**, *31*, e1–e5.
79. Pirotte, B.; Goldman, S.; David, P.; Wikler, D.; Damhaut, P.; Vandesteene, A.; Salmon, L.; Brotchi, J.; Levivier, M. Stereotactic brain biopsy guided by positron emission tomography (PET) with [F-18]fluorodeoxyglucose and [C-11]methionine. *Acta Neurochir. Suppl.* **1997**, *68*, 133–138. [PubMed]
80. Pirotte, B.; Goldman, S.; Massager, N.; David, P.; Wikler, D.; Vandesteene, A.; Salmon, I.; Brotchi, J.; Levivier, M. Comparison of 18F-FDG and 11C-methionine for PET-guided stereotactic brain biopsy of gliomas. *J. Nucl. Med. Off. Publ. Soc. Nucl. Med.* **2004**, *45*, 1293–1298.
81. Pafundi, D.H.; Laack, N.N.; Youland, R.S.; Parney, I.F.; Lowe, V.J.; Giannini, C.; Kemp, B.J.; Grams, M.P.; Morris, J.M.; Hoover, J.M.; et al. Biopsy validation of 18F-DOPA PET and biodistribution in gliomas for neurosurgical planning and radiotherapy target delineation: Results of a prospective pilot study. *Neuro-Oncology* **2013**, *15*, 1058–1067. [CrossRef] [PubMed]
82. Pirotte, B.J.; Lubansu, A.; Massager, N.; Wikler, D.; Goldman, S.; Levivier, M. Results of positron emission tomography guidance and reassessment of the utility of and indications for stereotactic biopsy in children with infiltrative brainstem tumors. *J. Neurosurg.* **2007**, *107* (Suppl. S5), 392–399. [CrossRef]
83. Galldiks, N.; Schroeter, M.; Fink, G.R.; Kracht, L.W. Interesting image. PET imaging of a butterfly glioblastoma. *Clin. Nucl. Med.* **2010**, *35*, 49–50. [CrossRef] [PubMed]
84. Calabria, F.; Schillaci, O. Recurrent glioma and crossed cerebellar diaschisis in a patient examined with 18F-DOPA and 18F-FDG PET/CT. *Clin. Nucl. Med.* **2012**, *37*, 878–879. [CrossRef] [PubMed]
85. Kebir, S.; Rauschenbach, L.; Gielen, G.H.; Schafer, N.; Tzaridis, T.; Scheffler, B.; Giordano, F.A.; Lazaridis, L.; Herrlinger, U.; Glas, M. Recurrent pseudoprogression in isocitrate dehydrogenase 1 mutant glioblastoma. *J. Clin. Neurosci. Off. J. Neurosurg. Soc. Australas* **2018**, *53*, 255–258. [CrossRef] [PubMed]
86. Peng, F.; Juhasz, C.; Bhambhani, K.; Wu, D.; Chugani, D.C.; Chugani, H.T. Assessment of progression and treatment response of optic pathway glioma with positron emission tomography using alpha-[(11)C]methyl-L-tryptophan. *Mol. Imaging Boil. MIB Off. Publ. Acad. Mol. Imaging* **2007**, *9*, 106–109. [CrossRef] [PubMed]
87. Viader, F.; Derlon, J.M.; Petit-Taboue, M.C.; Shishido, F.; Hubert, P.; Houtteville, J.P.; Courtheoux, P.; Chapon, F. Recurrent oligodendroglioma diagnosed with 11C-L-methionine and pet: A case report. *Eur. Neurol.* **1993**, *33*, 248–251. [CrossRef] [PubMed]
88. O'Halloran, P.J.; Viel, T.; Murray, D.W.; Wachsmuth, L.; Schwegmann, K.; Wagner, S.; Kopka, K.; Jarzabek, M.A.; Dicker, P.; Hermann, S.; et al. Mechanistic interrogation of combination bevacizumab/dual PI3K/mTOR inhibitor response in glioblastoma implementing novel MR and PET imaging biomarkers. *Eur. J. Nucl. Med. Mol. Imaging* **2016**, *43*, 1673–1683. [CrossRef] [PubMed]

89. Deuschl, C.; Moenninghoff, C.; Goericke, S.; Kirchner, J.; Koppen, S.; Binse, I.; Poeppel, T.D.; Quick, H.H.; Forsting, M.; Umutlu, L.; et al. Response assessment of bevacizumab therapy in GBM with integrated 11C-MET-PET/MRI: A feasibility study. *Eur. J. Nucl. Med. Mol. Imaging* **2017**, *44*, 1285–1295. [CrossRef] [PubMed]

90. Deuschl, C.; Goericke, S.; Grueneisen, J.; Sawicki, L.M.; GOEBEL, J.; El Hindy, N.; Wrede, K.; Binse, I.; Poeppel, T.; Quick, H.; et al. Simultaneous 11C-Methionine Positron Emission Tomography/Magnetic Resonance Imaging of Suspected Primary Brain Tumors. *PLoS ONE* **2016**, *11*, e0167596. [CrossRef] [PubMed]

91. Garcia, J.R.; Cozar, M.; Baquero, M.; Fernandez Barrionuevo, J.M.; Jaramillo, A.; Rubio, J.; Maida, G.; Soler, M.; Riera, E. The value of (11)C-methionine PET in the early differentiation between tumour recurrence and radionecrosis in patients treated for a high-grade glioma and indeterminate MRI. *Rev. Esp. Med. Nucl. Imagen Mol.* **2017**, *36*, 85–90. [PubMed]

92. Gauvain, K.; Ponisio, M.R.; Barone, A.; Grimaldi, M.; Parent, E.; Leeds, H.; Goyal, M.; Rubin, J.; McConathy, J. (18)F-FDOPA PET/MRI for monitoring early response to bevacizumab in children with recurrent brain tumors. *Neuro-Oncol. Pract.* **2018**, *5*, 28–36. [CrossRef] [PubMed]

93. Filss, C.P.; Cicone, F.; Shah, N.J.; Galldiks, N.; Langen, K.J. Amino acid PET and MR perfusion imaging in brain tumours. *Clin. Transl. Imaging* **2017**, *5*, 209–223. [CrossRef] [PubMed]

94. Muoio, B.; Giovanella, L.; Treglia, G. Recent Developments of 18F-FET PET in Neuro-oncology. *Curr. Med. Chem.* **2018**, *25*, 3061–3073. [CrossRef] [PubMed]

95. Albert, N.L.; Weller, M.; Suchorska, B.; Galldiks, N.; Soffietti, R.; Kim, M.M.; la Fougere, C.; Pope, W.; Law, I.; Arbizu, J.; et al. Response Assessment in Neuro-Oncology working group and European Association for Neuro-Oncology recommendations for the clinical use of PET imaging in gliomas. *Neuro-Oncology* **2016**, *18*, 1199–1208. [CrossRef] [PubMed]

© 2018 by the authors. Licensee MDPI, Basel, Switzerland. This article is an open access article distributed under the terms and conditions of the Creative Commons Attribution (CC BY) license (http://creativecommons.org/licenses/by/4.0/).

bioengineering

MDPI

Review

Radiotherapy Advances in Pediatric Neuro-Oncology

Ethan B. Ludmir, David R. Grosshans and Kristina D. Woodhouse *

Department of Radiation Oncology, Unit 97, The University of Texas MD Anderson Cancer Center,
1515 Holcombe Blvd, Houston, TX 77030, USA; ebludmir@mdanderson.org (E.B.L.);
dgrossha@mdanderson.org (D.R.G.)
* Correspondence: kdwoodhouse@mdanderson.org; Tel.: +1-713-563-2328

Received: 6 October 2018; Accepted: 1 November 2018; Published: 4 November 2018

Abstract: Radiation therapy (RT) represents an integral component in the treatment of many pediatric brain tumors. Multiple advances have emerged within pediatric radiation oncology that aim to optimize the therapeutic ratio—improving disease control while limiting RT-related toxicity. These include innovations in treatment planning with magnetic resonance imaging (MRI) simulation, as well as increasingly sophisticated radiation delivery techniques. Advanced RT techniques, including photon-based RT such as intensity-modulated RT (IMRT) and volumetric-modulated arc therapy (VMAT), as well as particle beam therapy and stereotactic RT, have afforded an array of options to dramatically reduce radiation exposure of uninvolved normal tissues while treating target volumes. Along with advances in image guidance of radiation treatments, novel RT approaches are being implemented in ongoing and future prospective clinical trials. As the era of molecular risk stratification unfolds, personalization of radiation dose, target, and technique holds the promise to meaningfully improve outcomes for pediatric neuro-oncology patients.

Keywords: radiotherapy; pediatric oncology; brain tumors; proton beam therapy; intensity-modulated radiotherapy; magnetic resonance imaging; image guidance

1. Introduction

Cancers of the central nervous system (CNS) are the most common solid tumors among pediatric patients, with an incidence of approximately 3000 new cases annually, accounting for 25% of all pediatric tumors [1]. Over the course of the last several decades, the prognosis for these malignancies has improved steadily, with the most recent nationwide data reporting an estimated 72% 5-year overall survival rate for the pooled cohort of pediatric patients with primary CNS tumors [1]. Improvements in the prognoses of these tumors have been made owing to a combination of factors, including advances in neurosurgery [2,3], diagnostic radiology [3–5], radiation oncology [6,7], and medical oncology [8,9], among others. In this review, we will provide an overarching view of recent advances in radiation oncology relevant to pediatric neuro-oncology, as well as look to the future of radiotherapy (RT) in the treatment of these challenging lesions.

2. Role of RT in Pediatric Neuro-Oncology

Pediatric brain tumors represent a heterogeneous group of lesions, whose clinical and biological characteristics vary widely; treatment approaches for these lesions can differ markedly between tumor types, patient demographics, and clinical context. Multimodality therapy for these tumors may consist of a combination of maximal safe neurosurgical resection, RT, and systemic chemotherapy or targeted agents. For many benign or low-grade lesions that are accessible surgically, neurosurgical resection alone may afford excellent outcomes, often with acceptable rates of toxicity [10]. On the other hand, for aggressive tumors, combination approaches including surgery, RT, and chemotherapy are indicated to achieve optimal outcomes [11].

When RT is indicated, the intent and target can vary depending on a number of factors as well. Among patients with embryonal brain tumors, such as those with medulloblastoma, RT is recommended in the adjuvant setting (following resection), generally targeting the entire brain and spine (craniospinal irradiation, CSI) as well as a boost to the tumor bed [12]. The aim of CSI in this context is to eradicate microscopic disease elsewhere in the CNS, particularly given the proclivity for embryonal tumors to disseminate via cerebrospinal fluid (CSF) pathways [12]. For patients with high-grade unresectable tumors, such as those with diffuse intrinsic pontine glioma (DIPG), RT remains *de facto* the sole therapeutic option available to delay progression and death. Thus, with a range of adjuvant, definitive, and palliative indications, RT plays a significant role in the treatment of most childhood brain tumors.

3. Advances in Treatment Planning

For the past two decades, RT treatment planning has relied on the use of computed tomography (CT) simulation. CT simulation allows providers, including the radiation oncologist as well as radiation therapist, to optimize the patient position for subsequent treatments. For patients undergoing RT to the brain, the simulation process often involves the manufacture and customization of a thermoplastic mask to immobilize the head. Following immobilization, axial CT images are acquired, transferred to the treatment planning system, and utilized to determine target volumes. These target volumes, including the gross tumor volume (GTV), clinical tumor volume (CTV), and planning target volume (PTV) are contoured onto the acquired CT simulation images. Also contoured are critical organs-at-risk or avoidance structures, including the brainstem, spinal cord, optic nerves, optic chiasm, cochleae, hippocampi, and others. Through delineation of these structures in the contouring process, treatment planning can proceed to optimize the dose to the target volumes while minimizing dose to critical normal structures.

One pitfall of this process in the treatment of brain tumors, however, is the relatively poor soft-tissue contrast associated with CT imaging, particularly in the neuro-axis. Therefore, from the initial development of CT simulation, efforts were made to incorporate the high degree of soft-tissue contrast with magnetic resonance imaging (MRI) into the treatment planning process [13–15]. To accomplish this, MRI sequences (generally from diagnostic scans) are co-registered to the CT simulation images. The co-registration, or 'fusion', of MRI series to CT imaging relies on transformation models to align these scans to one another, which can be optimized to account for organ distortion [16–18]. Multiple studies have demonstrated that the addition of MRI co-registration for treatment of brain tumors has resulted in substantial changes to both the contoured target volumes as well as organs-at-risk [15,19]. Variations in tumor delineation during treatment planning in the absence of co-registered MRI sequences carries significant clinical consequences, with the potential for under-treatment of disease as well as over-treatment of critical normal structures.

Given the centrality of MRI in treatment planning for brain tumors (among others), an increasing number of institutions are pursuing the use of dedicated MRI simulators [16,19]. These MRI simulators allow for MRI sequences to be obtained with appropriate patient positioning using requisite immobilization devices for treatment planning purposes (Figure 1). Whereas co-registered MRI data from diagnostic imaging are derived from scans in which patients are not in the treatment position or with immobilization devices such as a mask, dedicated MRI simulators allow for high-fidelity MRI sequences to be obtained in the treatment position [20]. This focus on geometric consistency between MRI acquisition and subsequent RT treatments facilitates utilization of MRI data for contour definition (Figure 1) [20,21]. Clinically, this approach is particularly advantageous in the context of adjacent target volumes and critical organs-at-risk, for instance treatment of posterior fossa tumors adjacent to brainstem, or suprasellar/sellar tumors adjacent to optic pathway structures; in these cases, accurate delineation of target versus non-target structures is paramount to avoid severe toxicity such as brainstem necrosis or visual impairment, respectively.

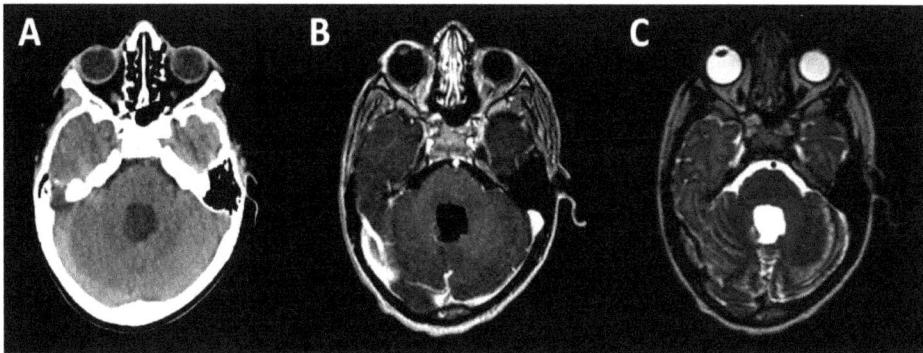

Figure 1. CT and MRI simulation. Axial images from CT (**A**) and MRI (**B,C**) simulation for radiation treatment planning purposes. This patient is an 8-year-old male who presents for adjuvant radiotherapy three weeks following gross total resection of average-risk posterior fossa medulloblastoma. CT simulator axial image showing the resection cavity is shown in (**A**); (**B,C**) show MRI simulator axial images from the same plane; T1 post-gadolinium contrast sequence is shown in (**B**), and the T2 sequence is shown in (**C**). The MRI simulator allows for identical positioning and immobilization to be achieved, facilitating fusion of CT and MRI images as demonstrated here.

Furthermore, efforts to date utilizing MRI co-registration or MRI simulation have generally used these as an adjunct to CT simulation. CT simulation provides a spatial electron density map (via CT-derived Hounsfield units), which is then used to model conformal RT treatment plans [22]. Modeling dose distribution accurately during treatment planning is contingent on this electron density map. Looking ahead, ongoing studies are attempting to utilize MRI sequences to infer electron density, so that accurate treatment planning could proceed with MRI simulation alone [16,21,23,24]. This would be particularly helpful for pediatric patients, in whom limiting radiation exposure (i.e., through reducing the number of CT scans performed) is of biological and clinical importance [25]. Logistically, pediatric patients up to the age of eight years old often require sedation to tolerate both simulation and radiation treatments; consolidating simulation to one scan (an MRI) may similarly provide practical advantages for this special patient population.

Specific MRI sequences and techniques also hold promise to enhance RT planning and delivery. Conventionally, commonly utilized MRI sequences to anatomically define target volumes as well as normal structures include T1 post-gadolinium-contrast and T2 fluid-attenuated inversion recovery (FLAIR) [16,26–28]. While T1 post-contrast and T2 FLAIR sequences yield excellent intracranial soft tissue contrast for defining volumes (T2 FLAIR being particularly helpful for assessing tissue peritumoral edema and gliosis), functional MRI sequences may expand the role of MRI in treatment planning as well as clinical outcomes [21,26]. Diffusion-weighted MRI sequences (also known as diffusion-weighted MR imaging, or DWI), for instance, provide an assessment of water diffusion through tissue, which can assist in assessing cellularity of a particular voxel [21]. Increased water diffusion distance reflects decreased cellularity—an indicator of treatment response. Supporting this, DWI mid-way through the course of RT has been shown among brain tumor patients to provide a radiographic biomarker of treatment response and subsequent long-term survival [29–31]. MR spectroscopy may provide further functional and metabolic data, using tissue metabolite nuclei characteristics to assess the chemical composition of a given voxel. With expertise and support from experienced radiologists, MR spectroscopy can yield a voxel-by-voxel map of data related to cell turnover, hypoxia, and other physiologic parameters that can inform treatment planning and response [21,26,32–34]. While promising, these functional imaging techniques are only now beginning to be incorporated into clinical practice.

4. Advances in RT Delivery

4.1. Advanced Photon RT Techniques

With the integration of CT simulation and three-dimensional treatment planning over the prior decades, photon-based RT techniques have similarly progressed. Moving beyond the era of two-dimensional (2D) and three-dimensional conformal RT (3D-CRT), intensity-modulated RT (IMRT) has entered mainstream clinical practice and use, including among pediatric neuro-oncology patients. The underlying premise for IMRT is rooted in the use of inverse-planning systems that rely on objective functions to optimize a treatment plan. Based on pre-specified target volume goals and organ-at-risk constraints, inverse-planning algorithms can generate plans using multiple beam arrangements (i.e., 8 or 9 beams) with non-uniform beam fluences to optimize the objective functions. These plans generally rely on the use of multileaf collimators (often made of a high-atomic-number [high-Z] materials such as tungsten) that subdivide each beam into several small 'beamlets', each of which in turn can have variable fluences to achieve the desired objective functions. Collectively, this approach can yield highly conformal plans in which high-dose regions can be shifted away from critical structures, though usually with a lower-dose 'bath' to a larger region of uninvolved tissue.

In the treatment of brain tumors, IMRT has been shown to improve target conformity as well as critical structure sparing when compared to standard 3D-CRT approaches [35–37]. These dosimetric advantages to IMRT are supported by clinical data, which demonstrate comparable rates of disease control with IMRT approaches [38–40], but decreased RT-related toxicity, particularly ototoxicity [37,41,42]. Expanding on IMRT, arc-based therapies have increasingly been utilized. These approaches, such as volumetric-modulated arc therapy (VMAT), rely on coplanar intensity-modulated arcs in which the linear accelerator gantry rotates around the patient [43,44]. As the gantry rotates around the patient, dynamic modulation can occur for the speed of gantry rotation, the beam-shaping aperture (including multi-leaf collimator), as well as the dose delivery rate [43–46]. The result is a highly-conformal treatment plan, comparable and often superior to IMRT with regard to target volume coverage, coverage homogeneity, and normal tissue sparing [44,45,47,48]. However, the primary advantage of arc-based techniques is the speed of treatment delivery [44], in which a single arc can be delivered over a few (2–3) min. Given that most VMAT plans rely on one to three arcs, the total time for patients on the treatment table may be reduced from approximately 30 min with IMRT to 10–15 min with VMAT, in our experience. This difference is particularly advantageous for pediatric patients, whose tolerance for treatments may be more limited than adult patients, and many of whom may require sedation while undergoing treatment. A caveat to the above is that the low-dose 'bath' may be more pronounced with VMAT, and certain clinical and dosimetric instances may arise where 3D-CRT approaches may provide optimal sparing of a specific at-risk structure [45,48].

4.2. Particle Therapy

In conjunction with advanced photon-based RT, particle therapy, including proton beam therapy (PBT), has now entered mainstream clinical practice among pediatric neuro-oncology patients [49–51]. The underlying principle of particle therapy, with PBT or heavier ions such as carbon ions, relies on the physical properties of particle beam dose deposition. With photon beams, radiation is deposited from its entrance into the body to a maximum dose, and then continues to deposit dose as it exits through to the other side of the body. The maximum depth dose for clinically-utilized photon beams (with energies between 4 MV and 18 MV) ranges between approximately 1.5 cm and 3.5 cm. Beyond these maximum depths, photon beams continue to deposit dose until exiting the other side of the body and striking a shielded wall on the opposite side of the treatment room. In contrast, proton beams (and more particle beams more generally) decrease velocity as they pass through tissue, depositing more energy with decreasing velocity until reaching a stopping depth in which the bulk of their energy is deposited. Distal to this characteristic depth, limited further energy is deposited in tissue, resulting in essentially no exit dose past the maximum depth dose for a proton beam. The characteristic peak energy deposition at a given depth is known as the Bragg curve (or Bragg peak). The specific depth of a Bragg peak for a mono-energetic proton beam is dependent on the energy of the proton beam (usually ranging between 70 and 250 MeV). By utilizing poly-energetic proton beams in clinical use, the Bragg peak becomes "spread-out" such that a target volume can be treated across its thickness with a proton beam; even with a poly-energetic proton beam, however, the dose drop-off at the end of the peak remains, with no dose beyond the distal edge of the spread-out Bragg peak. The primary advantage of PBT, therefore, is the ability to deliver RT with minimal exit dose.

Given the susceptibility of children to late effects from RT (including neurocognitive deficits, neuroendocrine deficits, audiovisual toxicity, growth abnormalities, and second malignancies, among others), the potential for PBT to minimize normal tissue radiation exposure through the absence of exit dose represents an enticing opportunity to enhance the therapeutic ratio. Multiple dosimetric studies have demonstrated a significant reduction in dose to critical organs at risk, as well as uninvolved brain more generally, in the treatment of diverse pediatric brain tumors using PBT; these studies confirmed dosimetric benefits of PBT over advanced photon-based techniques (including IMRT) for patients with medulloblastoma, ependymoma, and craniopharyngioma [52–57]. This has been most pronounced for patients who require CSI, such as medulloblastoma patients (Figure 2). With CSI, posterior positioning of proton beam(s) targeting the spinal canal as part of CSI allows for dose to cover the spinal canal and vertebral bodies, with minimal dose anteriorly into critical organs including thyroid, heart, lungs, gastrointestinal structures (i.e., pancreas), and ovaries, among others (Figure 2) [52,53,57,58]. So pronounced are the dosimetric differences in sparing anterior structures with PBT for CSI that the pediatric radiation oncology community has debated in recent years whether PBT represents the only standard of care option for pediatric patients requiring CSI (Figure 2) [59,60].

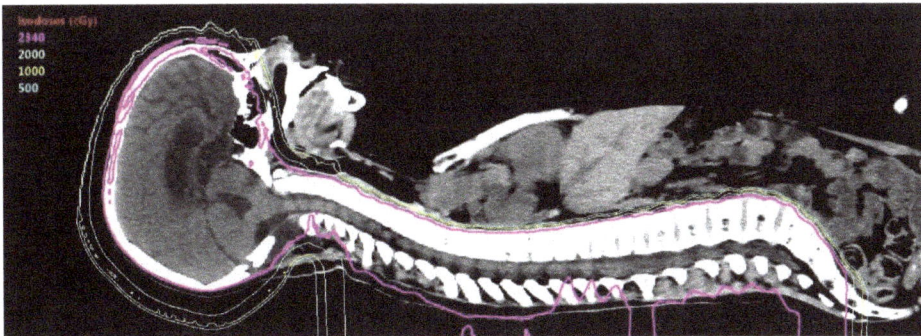

Figure 2. Proton beam therapy CSI. Representative mid-sagittal image of an 11-year-old male patient status post gross total resection for average-risk posterior-fossa medulloblastoma being planned for adjuvant radiotherapy (RT) including craniospinal irradiation (CSI). Shown here are isodose lines (colored lines) for the radiation treatment plan for the CSI component of this patient's adjuvant RT. CSI fields cover the complete brain and spinal cord to the termination of the thecal sac in the sacrum. The 100% isodose line (23.4Gy[RBE]) covers the entire target volume as shown in this image. As this patient is being treated with PBT, the posterior-positioned spinal fields have their terminal Bragg peak at the anterior-most portion of the vertebral bodies. Therefore, there is no exit dose anterior to the vertebrae, as shown with the isodose lines, including the low-dose 5Gy[RBE] isodose line (cyan). No dose is seen treating anterior structures such as the heart, liver, thyroid, gastrointestinal organs, or other anterior structures seen in this representative sagittal slice.

Dosimetric data have been supported by clinical data for PBT thus far. Across multiple disease sites, reviewed individually elsewhere, PBT has provided disease control rates comparable to those with photon-based RT, while often providing clinically-meaningful reductions in toxicity [7,49–51,58,61–66]. Furthermore, PBT does not appear to increase the risk of second malignancies among treated patients as compared with photon RT [67,68]. The possibility of increased second malignancy risk with PBT arose due to concerns regarding increased secondary neutron production with PBT, particularly with the use of patient-specific brass apertures and scattering devices utilized with certain PBT treatment plans [69,70]. Such secondary neutrons might, hypothetically, increase whole-body non-target radiation exposure and increase second malignancy risk [70], a relevant concern for pediatric patients whose baseline risk for RT-related second malignancy is higher than that of adults. However, both modeling data as well as clinical evidence to date suggests that if anything PBT may decrease second malignancy, likely owing to the absence of low-dose 'bath' as occurs with advanced photon-based techniques such as IMRT or VMAT [49,67,68,71].

The underlying technology and treatment delivery systems for PBT continue to progress as well. Earlier iterations of PBT have relied on the use of passive-scatter PBT (PSPT). PSPT utilizes a range modulator wheel to generate a spread-out Bragg peak from an initial monoenergetic proton beam; the proton beam is then shaped with brass apertures to sharpen the lateral border of the proton beam as well as compensator to modulate the distal edge of the beam. PSPT is somewhat limited with regard to modulating dose for structures proximal to the target volume in the path of the beam, and similarly may not provide conformal treatment plans for irregularly-shaped targets. Scanning-beam PBT, as opposed to PSPT, relies on proton 'beamlets' of discrete energies which can be delivered in layers (with each layer/depth reflecting a specific beamlet energy). Rather than using a range modulator, scanning-beam PBT allows for 'dose painting' of the target volume with these beamlets, and can afford high conformality of irregularly-shaped targets. Scanning-beam PBT therefore allows for intensity-modulated proton therapy (IMPT) to be possible, wherein 'dose painting' of a target volume can achieve highly-conformal dose distributions surpassing both advanced photon-based techniques as well as PSPT [72,73]. Emerging clinical data highlight high rates of disease control with reduced toxicity utilizing IMPT, primarily in the treatment of head and neck malignancies [74–76]. These advances in PBT technology, as well as challenges with regard to uncertainties with PBT, have been reviewed elsewhere and are beyond the scope of this review [49,51]. However, ongoing efforts to understand the radiobiological differences between proton and photon beams are increasingly revealing the complexities of particle therapy, as well as the enhanced biological effectiveness of protons as compared with photons [49]. The higher relative biological effectiveness of protons carries both potential advantages with regard to tumor cell killing, as well as risks in terms of increased risk of toxicity to critical structures that receive dose (such as the brainstem in the case of posterior fossa tumors) [49,77].

Heavier ion particle therapy, such as carbon ion radiotherapy, has also emerged as a treatment option for pediatric patients. Carbon ion therapy, as with PBT, demonstrates a Bragg peak dose distribution with minimal exit dose; however, the biological effectiveness of heavier ions is thought to be greater than that of both photons and protons [78]. These properties suggest that heavy ion therapy may be particularly helpful for radioresistant tumors, such as osteosarcoma (the most common pediatric bone tumor). Early data utilizing definitive heavy ion therapy for unresectable pediatric osteosarcoma are promising [79–81], and future studies exploring the role of heavy ion therapy in the treatment of pediatric patients are underway.

4.3. Stereotactic Approaches

Stereotactic RT represents another option emerging within pediatric radiation oncology. Stereotactic approaches are advanced photon-based techniques that are highly-focal and precise, delivering large fractional doses of radiation largely to small, spherical targets. Stereotactic RT is often delivered as a single fraction (generally called stereotactic radiosurgery [SRS] if delivered as a single treatment) or over a few fractions (usually five or fewer treatments, called stereotactic radiotherapy [SRT]). SRS and SRT can be delivered using a number of systems, including the Gamma Knife system (in which a patient sits inside a 'helmet' with multiple non-coplanar slits, which in turn allow for dose to be delivered from cobalt-60 sources to an intracranial target) as well as modified linear accelerators (including dedicated radiosurgery liner accelerators such as the CyberKnife system) [6]. The ideal target for such stereotactic approaches is small and spherical, and consequently this approach has been largely utilized for patients (primarily adults) with brain metastases. In the pediatric setting, stereotactic techniques present unique opportunities as well as challenges. While the proportion of pediatric patients with brain metastases is marginal as compared with that of adult patients, pediatric patients with both neoplastic as well as certain vascular lesions may benefit from SRS/SRT. Since SRS and SRT are delivered in one or a few fractions, respectively, this significantly reduces the burden of prolonged treatment duration on pediatric patients. On the other hand, as SRS and SRT approaches often require more invasive immobilization methods, such as the use of a stereotactic head 'frame' which is affixed to the skull with the use of metal pins, pediatric patients may be more likely to require sedation with the procedure. However, if the treatment is delivered only once (i.e., SRS) over a single day, this invasive but shorter approach may represent an advantage compared with daily sedation for a conventional RT course that would otherwise require 30 treatments over six weeks, for instance.

SRS and SRT themselves have been utilized heterogeneously among pediatric patients. This includes efforts to incorporate SRS as a 'boost' for high-risk ependymoma patients being treated with adjuvant RT [6,82,83], as definitive treatment for low-grade gliomas [84–86], or in the setting of recurrent/residual disease, often within previously-irradiated tissue [6,83,87–92]. One such case of utilizing SRS to treat a recurrent pituitary adenoma in a pediatric patient is highlighted in Figure 3. SRS has similarly been shown to represent a safe and effective treatment option for pediatric arteriovenous malformations, discussed and reviewed extensively elsewhere [93]. Beyond this, stereotactic approaches have been considered in a variety of clinical contexts (often borrowing upon experiences treating adult patients). Single-fraction high-dose palliative approaches with stereotactic RT are promising options, as is the use of stereotactic RT for the ablative treatment of oligometastatic disease [94,95]. Reirradiation may be another context where SRS/SRT may be helpful, given the rapid dose fall-off with such treatment plans and consequent minimal reirradiation of critical normal structures being exposed to another course of RT [96]. Stereotactic fractionated approaches are similarly considered for higher dose-per-fraction boost fields among certain pediatric ependymoma protocols [97]. While the indications for SRS and SRT among pediatric neuro-oncology patients remain heterogeneous, they represent an emerging technique for the treatment of intracranial lesions across diverse clinical contexts.

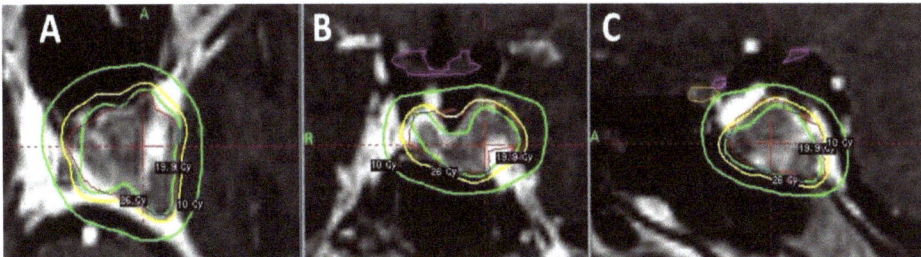

Figure 3. Stereotactic RT in the pediatric setting. This case is that of a 17-year-old female who presents with enlarging residual pituitary adenoma. The patient was initially treated one year prior with subtotal trans-nasal endoscopic resection of a non-secreting pituitary adenoma. On subsequent MRI surveillance, she was noted to have slow enlargement of residual pituitary adenoma. She was therefore recommended for salvage stereotactic RT to this residual disease. She was treated with a single-fraction of stereotactic radiosurgery, prescribing 19.9 Gy to the 50% isodose line. Shown here are representative (**A**) axial; (**B**) coronal; and (**C**) sagittal sections of the plan, with 10 Gy, 19.9 Gy, and 26 Gy isodose lines. Critical structures are appropriately avoided using this highly-conformal single-treatment technique, including optic chiasm (purple contour) and left optic nerve (orange contour).

4.4. Improving Image Guidance

Central to the use of advanced RT delivery technologies is improved patient alignment confirmation. For highly-conformal RT techniques, including IMRT, VMAT, PBT, and stereotactic RT, patient set-up errors and positional deviations can dramatically impact the treatment plan, potentially causing target volumes to be under-treated and normal structures to be over-treated. Such errors can therefore result in decreased tumor control, increased treatment-related toxicity, or both. Therefore, to support advanced RT delivery technologies, advanced alignment verification technologies have emerged. At the heart of these alignment verification methods are techniques known as image-guided radiation therapy (IGRT). IGRT most commonly includes techniques such as (1) kilovoltage (kV) planar radiographs from X-ray devices mounted in the treatment room (often mounted as part of the treatment linear accelerator/gantry itself), or (2) volumetric imaging using cone beam computed tomography (CBCT) scans, which are similarly incorporated into the treatment room [98,99]. kV-IGRT provides physicians with planar radiographs, often obtained in two radiographic planes (for instance, one anteroposterior radiograph and one lateral radiograph), that facilitate accurate alignment to bony landmarks [98]. CBCT, on the other hand, allows for three-dimensional volumetric imaging with CT-level soft tissue resolution [98]. For most intracranial tumors, our institutional practice, consistent with many other academic centers [98], has been to utilize daily kV-IGRT to ensure accurate bony alignment. Given the reliability of bony landmarks for intracranial target volumes, whose positions are markedly more consistent than targets in the abdomen or pelvis, for instance, daily kV imaging has been widely adopted as an effective IGRT tool for advanced RT techniques in the treatment of intracranial tumors [98,99]. Some institutions have more routinely employed CBCT imaging for pediatric neuro-oncology patients, suggesting that the set-up uncertainty margin added to the CTV to generate the PTV can be reduced with the use of CBCT [100]. However, concerns regarding radiation exposure to pediatric patients undergoing CBCT IGRT have been raised [25,99]. With doses of approximately 3 cGy per CBCT [101], frequent CBCT imaging (daily or weekly) has the potential to substantially increase risk of mutagenesis and second malignancy [25,99]. This additional radiation exposure may also translate into higher rates of infertility and endocrine dysfunction [25,99]. Therefore, our overarching approach, consistent with most high-volume pediatric centers, has been to utilize kV-IGRT to optimize patient alignment and reduce PTV margins; CBCT, however, is used sparingly.

Along these lines, an important caveat is that IGRT as well as adaptive re-planning have been increasingly utilized for target volumes whose size and shape may vary over the treatment course. Perhaps the clearest example of such a tumor within pediatric neuro-oncology is craniopharyngioma. These lesions, comprised of both solid and cystic components, have a propensity to develop dynamic changes in cyst size and position during a course of RT [102,103]. Therefore, adaptive re-planning with repeat imaging has been recommended by multiple groups in the treatment of craniopharyngioma [55,62,102–104]. Our practice is generally to have craniopharyngioma patients undergo MRI simulation up-front as well as every one to two weeks while on treatment. Using the MRI simulator, we can obtain MRI imaging to assess cyst dynamics and complete an adaptive radiation plan if needed utilizing these MR images in the treatment position. This paradigm also avoids repeat CT imaging and associated radiation exposure, ideal for pediatric patients as previously discussed.

5. Clinical Efforts—Refining Target Volumes

5.1. Shrinking Field Sizes

Cooperative group clinical studies have further advanced pediatric neuro-oncology with efforts to refine target volumes. By decreasing field sizes, many of these trials aim to reduce RT-related toxicity and exposure without compromising disease control. For medulloblastoma patients, institutional data suggested that the standard boost field encompassing the entire posterior fossa (PF) may be excessive, and that rather a smaller 'involved field' (IF) boost volume targeting the tumor bed (with margin) would be sufficient [105,106]. To test whether such a reduced IF boost would provide comparable disease control rates as a PF boost, the Children's Oncology Group (COG) ACNS 0331 trial randomized standard-risk medulloblastoma patients to IF versus PF boost fields; presented in abstract form, the data demonstrate comparable 5-year event free survival (EFS) between the IF and PF boost arms (82.2% vs. 80.8%) [107]. Consequently, the standard of care has now shifted supporting use of IF boost fields rather than PF boost fields for medulloblastoma patients. Along similar lines, intracranial germ cell tumor (GCT) studies have aimed to reduce field sizes. For intracranial pure germinomas, a highly radiosensitive tumor, efforts sought to reduce definitive RT volumes from CSI to limited-field RT targeting the local tumor alone [108,109]. These prospective European studies demonstrated that such drastic target volume reductions yielded unacceptably high rates of periventricular disease relapse [108,109]. These data, coupled with a large literature review, supported the use of whole-ventricular irradiation (WVI), an intermediate between CSI and limited-field RT [108–110]. WVI, while still somewhat larger than limited-field RT targeting the tumor alone, represents a significant reduction in the treatment volume as compared with CSI, and has emerged as the standard for intracranial germinoma, utilized in the most recent COG protocol (ACNS 1123) [111]. For non-germinomatous intracranial germ cell tumors (NGGCTs), which have a higher risk of relapse and a poorer prognosis as compared with pure germinomas, the North American treatment paradigm has centered on the use of induction chemotherapy, with or without second-look surgery, and subsequent CSI [112]. As with medulloblastoma, institutional data suggest that reduced field size to WVI may not compromise disease control [113]; the NGGCT stratum of ACNS 1123 therefore is testing whether WVI, rather than CSI, may be an effective option for those who radiographically respond to induction chemotherapy.

5.2. Optimizing Dose

In conjunction with prospective studies designed to reduce field size, pediatric neuro-oncology trials have also focused on better elucidating the optimal RT doses for patients. These protocols have broadly asked to what extent RT dose can be reduced without compromising clinical outcomes. For standard-risk medulloblastoma patients, for instance, prospective data demonstrated that adjuvant RT with dose-reduced CSI (from 36 Gy to 23.4 Gy) did not result in inferior outcomes, provided patients also received adjuvant chemotherapy following RT [114]. COG ACNS 0331 attempted to further reduce the CSI dose to 18 Gy for standard-risk medulloblastoma patients, based on promising pilot data [107,115]. However, despite these pilot studies, reduced dose CSI from 23.4 Gy to 18 Gy in ACNS 0331 resulted in inferior EFS and overall survival [107]. Therefore, these trials have established 23.4 Gy, and not lower, as the ideal CSI dose for standard-risk medulloblastoma patients [107]. Efforts to determine the appropriate RT dose for pediatric patients are increasingly incorporating radiographic and molecular data to better risk-stratify patients and tailor dose accordingly. For intracranial germinoma patients, ACNS 1123 is assessing the feasibility of dose reduction based on radiographic response to induction chemotherapy [111]. For medulloblastoma patients, ongoing trials in both North America (SJMB12) and Europe (PNET5) are utilizing medulloblastoma molecular risk stratification to tailor RT dose, including dose reduction for lower-risk patients as well as dose intensification for higher-risk patients [116].

6. Conclusions

The treatment of pediatric brain tumors with RT continues to evolve with ongoing advances. New technologies, from MRI simulation to particle beam therapy and stereotactic RT, aim to improve the therapeutic ratio further—optimizing disease control and minimizing treatment-related toxicity. As the era of molecular risk stratification unfolds, personalization of radiation dose, target, and technique holds the promise to better the lives of all pediatric neuro-oncology patients.

Author Contributions: Conceptualization, E.B.L., D.R.G, and K.D.W.; Writing-Original Draft Preparation, E.B.L.; Writing-Review & Editing, E.B.L., D.R.G., and K.D.W.

Bioengineering 2018, 5, 97

Funding: This research received no external funding.

Conflicts of Interest: All authors report no conflicts of interest or relevant financial disclosures related to this work.

References

1. Ward, E.; DeSantis, C.; Robbins, A.; Kohler, B.; Jemal, A. Childhood and adolescent cancer statistics, 2014. *CA A Cancer J. Clin.* **2014**, *64*, 83–103. [CrossRef] [PubMed]
2. Governale, L.S. Minimally invasive pediatric neurosurgery. *Pediatr. Neurol.* **2015**, *52*, 389–397. [CrossRef] [PubMed]
3. Zebian, B.; Vergani, F.; Lavrador, J.P.; Mukherjee, S.; Kitchen, W.J.; Stagno, V.; Chamilos, C.; Pettorini, B.; Mallucci, C. Recent technological advances in pediatric brain tumor surgery. *CNS Oncol.* **2017**, *6*, 71–82. [CrossRef] [PubMed]
4. Choudhri, A.F.; Siddiqui, A.; Klimo, P., Jr.; Boop, F.A. Intraoperative MRI in pediatric brain tumors. *Pediatr. Radiol.* **2015**, *45* (Suppl. 3), S397–S405. [CrossRef] [PubMed]
5. Choudhri, A.F.; Siddiqui, A.; Klimo, P., Jr. Pediatric Cerebellar Tumors: Emerging Imaging Techniques and Advances in Understanding of Genetic Features. *Magn. Reson. Imaging Clin. N. Am.* **2016**, *24*, 811–821. [CrossRef] [PubMed]
6. Murphy, E.S.; Chao, S.T.; Angelov, L.; Vogelbaum, M.A.; Barnett, G.; Jung, E.; Recinos, V.R.; Mohammadi, A.; Suh, J.H. Radiosurgery for Pediatric Brain Tumors. *Pediatr. Blood Cancer* **2016**, *63*, 398–405. [CrossRef] [PubMed]
7. Eaton, B.R.; Yock, T. The use of proton therapy in the treatment of benign or low-grade pediatric brain tumors. *Cancer J.* **2014**, *20*, 403–408. [CrossRef] [PubMed]
8. Gajjar, A.; Pfister, S.M.; Taylor, M.D.; Gilbertson, R.J. Molecular insights into pediatric brain tumors have the potential to transform therapy. *Clin. Cancer Res. Off. J. Am. Assoc. Cancer Res.* **2014**, *20*, 5630–5640. [CrossRef] [PubMed]
9. Kieran, M.W. Targeting BRAF in pediatric brain tumors. *Am. Soc. Clin. Oncol. Educ. Book* **2014**. [CrossRef] [PubMed]
10. Watson, G.A.; Kadota, R.P.; Wisoff, J.H. Multidisciplinary management of pediatric low-grade gliomas. *Semin. Radiat. Oncol.* **2001**, *11*, 152–162. [CrossRef] [PubMed]
11. Paulino, A.C. Current multimodality management of medulloblastoma. *Curr. Prob. Cancer* **2002**, *26*, 317–356. [CrossRef]
12. McGovern, S.L.; Grosshans, D.; Mahajan, A. Embryonal brain tumors. *Cancer J.* **2014**, *20*, 397–402. [CrossRef] [PubMed]
13. Just, M.; Rosler, H.P.; Higer, H.P.; Kutzner, J.; Thelen, M. MRI-assisted radiation therapy planning of brain tumors–clinical experiences in 17 patients. *Magn. Reson. Imaging* **1991**, *9*, 173–177. [CrossRef]
14. Ten Haken, R.K.; Thornton, A.F., Jr.; Sandler, H.M.; LaVigne, M.L.; Quint, D.J.; Fraass, B.A.; Kessler, M.L.; McShan, D.L. A quantitative assessment of the addition of MRI to CT-based, 3-D treatment planning of brain tumors. *Radiother. Oncol. J. Eur. Soc. Ther. Radiol. Oncol.* **1992**, *25*, 121–133. [CrossRef]
15. Thornton, A.F., Jr.; Sandler, H.M.; Ten Haken, R.K.; McShan, D.L.; Fraass, B.A.; La Vigne, M.L.; Yanke, B.R. The clinical utility of magnetic resonance imaging in 3-dimensional treatment planning of brain neoplasms. *Int. J. Radiat. Oncol. Biol. Phys.* **1992**, *24*, 767–775. [CrossRef]
16. Devic, S. MRI simulation for radiotherapy treatment planning. *Med. Phys.* **2012**, *39*, 6701–6711. [CrossRef] [PubMed]
17. Brock, K.K. Results of a multi-institution deformable registration accuracy study (MIDRAS). *Int. J. Radiat. Oncol. Biol. Phys.* **2010**, *76*, 583–596. [CrossRef] [PubMed]
18. Wu, X.; Dibiase, S.J.; Gullapalli, R.; Yu, C.X. Deformable image registration for the use of magnetic resonance spectroscopy in prostate treatment planning. *Int. J. Radiat. Oncol. Biol. Phys.* **2004**, *58*, 1577–1583. [CrossRef] [PubMed]
19. Weber, D.C.; Wang, H.; Albrecht, S.; Ozsahin, M.; Tkachuk, E.; Rouzaud, M.; Nouet, P.; Dipasquale, G. Open low-field magnetic resonance imaging for target definition, dose calculations and set-up verification during three-dimensional CRT for glioblastoma multiforme. *Clin. Oncol. (R. Coll. Radiol.)* **2008**, *20*, 157–167. [CrossRef] [PubMed]

20. Rai, R.; Kumar, S.; Batumalai, V.; Elwadia, D.; Ohanessian, L.; Juresic, E.; Cassapi, L.; Vinod, S.K.; Holloway, L.; Keall, P.J.; et al. The integration of MRI in radiation therapy: Collaboration of radiographers and radiation therapists. *J. Med. Radiat. Sci.* **2017**, *64*, 61–68. [CrossRef] [PubMed]

21. Metcalfe, P.; Liney, G.P.; Holloway, L.; Walker, A.; Barton, M.; Delaney, G.P.; Vinod, S.; Tome, W. The potential for an enhanced role for MRI in radiation-therapy treatment planning. *Technol. Cancer Res. Treat.* **2013**, *12*, 429–446. [CrossRef] [PubMed]

22. Matsufuji, N.; Tomura, H.; Futami, Y.; Yamashita, H.; Higashi, A.; Minohara, S.; Endo, M.; Kanai, T. Relationship between CT number and electron density, scatter angle and nuclear reaction for hadron-therapy treatment planning. *Phys. Med. Biol.* **1998**, *43*, 3261–3275. [CrossRef] [PubMed]

23. Scheffler, K.; Lehnhardt, S. Principles and applications of balanced SSFP techniques. *European radiology* **2003**, *13*, 2409–2418. [CrossRef] [PubMed]

24. Haase, A.; Frahm, J.; Matthaei, D.; Hanicke, W.; Merboldt, K.D. FLASH imaging: Rapid NMR imaging using low flip-angle pulses. 1986. *J. Magn. Reson.* **2011**, *213*, 533–541. [CrossRef] [PubMed]

25. Brenner, D.J.; Hall, E.J. Computed tomography–an increasing source of radiation exposure. *N. Engl. J. Med.* **2007**, *357*, 2277–2284. [CrossRef] [PubMed]

26. Liney, G.P.; Moerland, M.A. Magnetic resonance imaging acquisition techniques for radiotherapy planning. *Semin. Radiat. Oncol.* **2014**, *24*, 160–168. [CrossRef] [PubMed]

27. Mazzara, G.P.; Velthuizen, R.P.; Pearlman, J.L.; Greenberg, H.M.; Wagner, H. Brain tumor target volume determination for radiation treatment planning through automated MRI segmentation. *Int. J. Radiat. Oncol. Biol. Phys.* **2004**, *59*, 300–312. [CrossRef] [PubMed]

28. Schad, L.R.; Bluml, S.; Hawighorst, H.; Wenz, F.; Lorenz, W.J. Radiosurgical treatment planning of brain metastases based on a fast, three-dimensional MR imaging technique. *Magn. Reson. Imaging* **1994**, *12*, 811–819. [CrossRef]

29. Mardor, Y.; Pfeffer, R.; Spiegelmann, R.; Roth, Y.; Maier, S.E.; Nissim, O.; Berger, R.; Glicksman, A.; Baram, J.; Orenstein, A.; et al. Early detection of response to radiation therapy in patients with brain malignancies using conventional and high b-value diffusion-weighted magnetic resonance imaging. *J. Clin. Oncol.* **2003**, *21*, 1094–1100. [CrossRef] [PubMed]

30. Hamstra, D.A.; Galban, C.J.; Meyer, C.R.; Johnson, T.D.; Sundgren, P.C.; Tsien, C.; Lawrence, T.S.; Junck, L.; Ross, D.J.; Rehemtulla, A.; et al. Functional diffusion map as an early imaging biomarker for high-grade glioma: Correlation with conventional radiologic response and overall survival. *J. Clin. Oncol.* **2008**, *26*, 3387–3394. [CrossRef] [PubMed]

31. Goldman, M.; Boxerman, J.L.; Rogg, J.M.; Noren, G. Utility of apparent diffusion coefficient in predicting the outcome of Gamma Knife-treated brain metastases prior to changes in tumor volume: A preliminary study. *J. Neurosurg.* **2006**, *105*, 175–182. [PubMed]

32. Hoskin, P.J.; Carnell, D.M.; Taylor, N.J.; Smith, R.E.; Stirling, J.J.; Daley, F.M.; Saunders, M.I.; Bentzen, S.M.; Collins, D.J.; d'Arcy, J.A.; et al. Hypoxia in Prostate Cancer: Correlation of BOLD-MRI With Pimonidazole Immunohistochemistry—Initial Observations. *Int. J. Radiat. Oncol. Biol. Phys.* **2007**, *68*, 1065–1071. [CrossRef] [PubMed]

33. Payne, G.S.; Leach, M.O. Applications of magnetic resonance spectroscopy in radiotherapy treatment planning. *Br. J. Radiol.* **2006**, *79*, S16–S26. [CrossRef] [PubMed]

34. Chang, J.; Thakur, S.; Perera, G.; Kowalski, A.; Huang, W.; Karimi, S.; Hunt, M.; Koutcher, J.; Fuks, Z.; Amols, H.; et al. Image-fusion of MR spectroscopic images for treatment planning of gliomas. *Med. Phys.* **2006**, *33*, 32–40. [CrossRef] [PubMed]

35. Hermanto, U.; Frija, E.K.; Lii, M.J.; Chang, E.L.; Mahajan, A.; Woo, S.Y. Intensity-modulated radiotherapy (IMRT) and conventional three-dimensional conformal radiotherapy for high-grade gliomas: Does IMRT increase the integral dose to normal brain? *Int. J. Radiat. Oncol. Biol. Phys.* **2007**, *67*, 1135–1144. [CrossRef] [PubMed]

36. Beltran, C.; Naik, M.; Merchant, T.E. Dosimetric effect of setup motion and target volume margin reduction in pediatric ependymoma. *Radiother. Oncol. J. Eur. Soc. Ther. Radiol. Oncol.* **2010**, *96*, 216–222. [CrossRef] [PubMed]

37. Huang, E.; Teh, B.S.; Strother, D.R.; Davis, Q.G.; Chiu, J.K.; Lu, H.H.; Carpenter, L.S.; Mai, W.Y.; Chintagumpala, M.M.; South, M.; et al. Intensity-modulated radiation therapy for pediatric medulloblastoma: Early report on the reduction of ototoxicity. *Int. J. Radiat. Oncol. Biol. Phys.* **2002**, *52*, 599–605. [CrossRef]

38. Paulino, A.C.; Mazloom, A.; Terashima, K.; Su, J.; Adesina, A.M.; Okcu, M.F.; Teh, B.S.; Chintagumpala, M. Intensity-modulated radiotherapy (IMRT) in pediatric low-grade glioma. *Cancer* **2013**, *119*, 2654–2659. [CrossRef] [PubMed]

39. Polkinghorn, W.R.; Dunkel, I.J.; Souweidane, M.M.; Khakoo, Y.; Lyden, D.C.; Gilheeney, S.W.; Becher, O.J.; Budnick, A.S.; Wolden, S.L. Disease control and ototoxicity using intensity-modulated radiation therapy tumor-bed boost for medulloblastoma. *Int. J. Radiat. Oncol. Biol. Phys.* **2011**, *81*, e15–e20. [CrossRef] [PubMed]

40. Greenfield, B.J.; Okcu, M.F.; Baxter, P.A.; Chintagumpala, M.; Teh, B.S.; Dauser, R.C.; Su, J.; Desai, S.S.; Paulino, A.C. Long-term disease control and toxicity outcomes following surgery and intensity modulated radiation therapy (IMRT) in pediatric craniopharyngioma. *Radiother. Oncol. J. Eur. Soc. Ther. Radiol. Oncol.* **2015**, *114*, 224–229. [CrossRef] [PubMed]

41. Nanda, R.H.; Ganju, R.G.; Schreibmann, E.; Chen, Z.; Zhang, C.; Jegadeesh, N.; Cassidy, R.; Deng, C.; Eaton, B.R.; Esiashvili, N. Correlation of Acute and Late Brainstem Toxicities With Dose-Volume Data for Pediatric Patients With Posterior Fossa Malignancies. *Int. J. Radiat. Oncol. Biol. Phys.* **2017**, *98*, 360–366. [CrossRef] [PubMed]

42. Jain, N.; Krull, K.R.; Brouwers, P.; Chintagumpala, M.M.; Woo, S.Y. Neuropsychological outcome following intensity-modulated radiation therapy for pediatric medulloblastoma. *Pediatr. Blood Cancer* **2008**, *51*, 275–279. [CrossRef] [PubMed]

43. Otto, K. Volumetric modulated arc therapy: IMRT in a single gantry arc. *Med. Phys.* **2008**, *35*, 310–317. [CrossRef] [PubMed]

44. Fogliata, A.; Clivio, A.; Nicolini, G.; Vanetti, E.; Cozzi, L. Intensity modulation with photons for benign intracranial tumours: A planning comparison of volumetric single arc, helical arc and fixed gantry techniques. *Radiother. Oncol. J. Eur. Soc. Ther. Radiol. Oncol.* **2008**, *89*, 254–262. [CrossRef] [PubMed]

45. Wagner, D.; Christiansen, H.; Wolff, H.; Vorwerk, H. Radiotherapy of malignant gliomas: Comparison of volumetric single arc technique (RapidArc), dynamic intensity-modulated technique and 3D conformal technique. *Radiother. Oncol. J. Eur. Soc. Ther. Radiol. Oncol.* **2009**, *93*, 593–596. [CrossRef] [PubMed]

46. Bush, K.; Townson, R.; Zavgorodni, S. Monte Carlo simulation of RapidArc radiotherapy delivery. *Phys. Med. Biol.* **2008**, *53*, N359–N370. [CrossRef] [PubMed]

47. Beltran, C.; Gray, J.; Merchant, T.E. Intensity-modulated arc therapy for pediatric posterior fossa tumors. *Int. J. Radiat. Oncol. Biol. Phys.* **2012**, *82*, e299–e304. [CrossRef] [PubMed]

48. Shaffer, R.; Nichol, A.M.; Vollans, E.; Fong, M.; Nakano, S.; Moiseenko, V.; Schmuland, M.; Ma, R.; McKenzie, M.; Otto, K. A comparison of volumetric modulated arc therapy and conventional intensity-modulated radiotherapy for frontal and temporal high-grade gliomas. *Int. J. Radiat. Oncol. Biol. Phys.* **2010**, *76*, 1177–1184. [CrossRef] [PubMed]

49. Mohan, R.; Grosshans, D. Proton therapy—Present and future. *Pediatr. Blood Cancer* **2017**, *109*, 26–44. [CrossRef] [PubMed]

50. Ladra, M.M.; MacDonald, S.M.; Terezakis, S.A. Proton therapy for central nervous system tumors in children. *Pediatr. Blood Cancer* **2018**, *65*, e27046. [CrossRef] [PubMed]

51. Chhabra, A.; Mahajan, A. Treatment of common pediatric CNS malignancies with proton therapy. *Chin. Clin. Oncol.* **2016**, *5*, 49. [CrossRef] [PubMed]

52. Jimenez, R.B.; Sethi, R.; Depauw, N.; Pulsifer, M.B.; Adams, J.; McBride, S.M.; Ebb, D.; Fullerton, B.C.; Tarbell, N.J.; Yock, T.I.; et al. Proton radiation therapy for pediatric medulloblastoma and supratentorial primitive neuroectodermal tumors: Outcomes for very young children treated with upfront chemotherapy. *Int. J. Radiat. Oncol. Biol. Phys.* **2013**, *87*, 120–126. [CrossRef] [PubMed]

53. St Clair, W.H.; Adams, J.A.; Bues, M.; Fullerton, B.C.; La Shell, S.; Kooy, H.M.; Loeffler, J.S.; Tarbell, N.J. Advantage of protons compared to conventional X-ray or IMRT in the treatment of a pediatric patient with medulloblastoma. *Int. J. Radiat. Oncol. Biol. Phys.* **2004**, *58*, 727–734. [CrossRef]

54. Boehling, N.S.; Grosshans, D.R.; Bluett, J.B.; Palmer, M.T.; Song, X.; Amos, R.A.; Sahoo, N.; Meyer, J.J.; Mahajan, A.; Woo, S.Y. Dosimetric comparison of three-dimensional conformal proton radiotherapy, intensity-modulated proton therapy, and intensity-modulated radiotherapy for treatment of pediatric craniopharyngiomas. *Int. J. Radiat. Oncol. Biol. Phys.* **2012**, *82*, 643–652. [CrossRef] [PubMed]

55. Beltran, C.; Roca, M.; Merchant, T.E. On the benefits and risks of proton therapy in pediatric craniopharyngioma. *Int. J. Radiat. Oncol. Biol. Phys.* **2012**, *82*, e281–e287. [CrossRef] [PubMed]

56. MacDonald, S.M.; Safai, S.; Trofimov, A.; Wolfgang, J.; Fullerton, B.; Yeap, B.Y.; Bortfeld, T.; Tarbell, N.J.; Yock, T. Proton radiotherapy for childhood ependymoma: Initial clinical outcomes and dose comparisons. *Int. J. Radiat. Oncol. Biol. Phys.* **2008**, *71*, 979–986. [CrossRef] [PubMed]

57. Brower, J.V.; Gans, S.; Hartsell, W.F.; Goldman, S.; Fangusaro, J.R.; Patel, N.; Lulla, R.R.; Smiley, N.P.; Chang, J.H.; Gondi, V. Proton therapy and helical tomotherapy result in reduced dose deposition to the pancreas in the setting of cranio-spinal irradiation for medulloblastoma: Implications for reduced risk of diabetes mellitus in long-term survivors. *Acta Oncol.* **2015**, *54*, 563–566. [CrossRef] [PubMed]

58. Eaton, B.R.; Esiashvili, N.; Kim, S.; Patterson, B.; Weyman, E.A.; Thornton, L.T.; Mazewski, C.; MacDonald, T.J.; Ebb, D.; MacDonald, S.M.; et al. Endocrine outcomes with proton and photon radiotherapy for standard risk medulloblastoma. *Neuro-Oncology* **2016**, *18*, 881–887. [CrossRef] [PubMed]

59. Wolden, S.L. Protons for craniospinal radiation: Are clinical data important? *Int. J. Radiat. Oncol. Biol. Phys.* **2013**, *87*, 231–232. [CrossRef] [PubMed]

60. Johnstone, P.A.; McMullen, K.P.; Buchsbaum, J.C.; Douglas, J.G.; Helft, P. Pediatric CSI: Are protons the only ethical approach? *Int. J. Radiat. Oncol. Biol. Phys.* **2013**, *87*, 228–230. [CrossRef] [PubMed]

61. Ladra, M.M.; Szymonifka, J.D.; Mahajan, A.; Friedmann, A.M.; Yong Yeap, B.; Goebel, C.P.; MacDonald, S.M.; Grosshans, D.R.; Rodriguez-Galindo, C.; Marcus, K.J.; et al. Preliminary results of a phase II trial of proton radiotherapy for pediatric rhabdomyosarcoma. *J. Clin. Oncol.* **2014**, *32*, 3762–3770. [CrossRef] [PubMed]

62. Bishop, A.J.; Greenfield, B.; Mahajan, A.; Paulino, A.C.; Okcu, M.F.; Allen, P.K.; Chintagumpala, M.; Kahalley, L.S.; McAleer, M.F.; McGovern, S.L.; et al. Proton beam therapy versus conformal photon radiation therapy for childhood craniopharyngioma: Multi-institutional analysis of outcomes, cyst dynamics, and toxicity. *Int. J. Radiat. Oncol. Biol. Phys.* **2014**, *90*, 354–361. [CrossRef] [PubMed]

63. McGovern, S.L.; Okcu, M.F.; Munsell, M.F.; Kumbalasseriyil, N.; Grosshans, D.R.; McAleer, M.F.; Chintagumpala, M.; Khatua, S.; Mahajan, A. Outcomes and acute toxicities of proton therapy for pediatric atypical teratoid/rhabdoid tumor of the central nervous system. *Int. J. Radiat. Oncol. Biol. Phys.* **2014**, *90*, 1143–1152. [CrossRef] [PubMed]

64. Sethi, R.V.; Giantsoudi, D.; Raiford, M.; Malhi, I.; Niemierko, A.; Rapalino, O.; Caruso, P.; Yock, T.I.; Tarbell, N.J.; Paganetti, H.; et al. Patterns of failure after proton therapy in medulloblastoma; linear energy transfer distributions and relative biological effectiveness associations for relapses. *Int. J. Radiat. Oncol. Biol. Phys.* **2014**, *88*, 655–663. [CrossRef] [PubMed]

65. Sato, M.; Gunther, J.R.; Mahajan, A.; Jo, E.; Paulino, A.C.; Adesina, A.M.; Jones, J.Y.; Ketonen, L.M.; Su, J.M.; Okcu, M.F.; et al. Progression-free survival of children with localized ependymoma treated with intensity-modulated radiation therapy or proton-beam radiation therapy. *Cancer* **2017**, *123*, 2570–2578. [CrossRef] [PubMed]

66. Gunther, J.R.; Sato, M.; Chintagumpala, M.; Ketonen, L.; Jones, J.Y.; Allen, P.K.; Paulino, A.C.; Okcu, M.F.; Su, J.M.; Weinberg, J.; et al. Imaging Changes in Pediatric Intracranial Ependymoma Patients Treated With Proton Beam Radiation Therapy Compared to Intensity Modulated Radiation Therapy. *Int. J. Radiat. Oncol. Biol. Phys.* **2015**, *93*, 54–63. [CrossRef] [PubMed]

67. Sethi, R.V.; Shih, H.A.; Yeap, B.Y.; Mouw, K.W.; Petersen, R.; Kim, D.Y.; Munzenrider, J.E.; Grabowski, E.; Rodriguez-Galindo, C.; Yock, T.I.; et al. Second nonocular tumors among survivors of retinoblastoma treated with contemporary photon and proton radiotherapy. *Cancer* **2014**, *120*, 126–133. [CrossRef] [PubMed]

68. Chung, C.S.; Yock, T.I.; Nelson, K.; Xu, Y.; Keating, N.L.; Tarbell, N.J. Incidence of second malignancies among patients treated with proton versus photon radiation. *Int. J. Radiat. Oncol. Biol. Phys.* **2013**, *87*, 46–52. [CrossRef] [PubMed]

69. Geng, C.; Moteabbed, M.; Xie, Y.; Schuemann, J.; Yock, T.; Paganetti, H. Assessing the radiation-induced second cancer risk in proton therapy for pediatric brain tumors: The impact of employing a patient-specific aperture in pencil beam scanning. *Phys. Med. Biol.* **2016**, *61*, 12–22. [CrossRef] [PubMed]

70. Brenner, D.J.; Hall, E.J. Secondary neutrons in clinical proton radiotherapy: A charged issue. *Radiother. Oncol. J. Eur. Soc. Ther. Radiol. Oncol.* **2008**, *86*, 165–170. [CrossRef] [PubMed]

71. Moteabbed, M.; Yock, T.I.; Paganetti, H. The risk of radiation-induced second cancers in the high to medium dose region: A comparison between passive and scanned proton therapy, IMRT and VMAT for pediatric patients with brain tumors. *Phys. Med. Biol.* **2014**, *59*, 2883–2899. [CrossRef] [PubMed]

72. Lomax, A. Intensity modulation methods for proton radiotherapy. *Phys. Med. Biol.* **1999**, *44*, 185–205. [CrossRef] [PubMed]

73. Lomax, A.J.; Boehringer, T.; Coray, A.; Egger, E.; Goitein, G.; Grossmann, M.; Juelke, P.; Lin, S.; Pedroni, E.; Rohrer, B.; et al. Intensity modulated proton therapy: A clinical example. *Med. Phy.* **2001**, *28*, 317–324. [CrossRef] [PubMed]

74. Frank, S.J.; Cox, J.D.; Gillin, M.; Mohan, R.; Garden, A.S.; Rosenthal, D.I.; Gunn, G.B.; Weber, R.S.; Kies, M.S.; Lewin, J.S.; et al. Multifield optimization intensity modulated proton therapy for head and neck tumors: A translation to practice. *Int. J. Radiat. Oncol. Biol. Phys.* **2014**, *89*, 846–853. [CrossRef] [PubMed]

75. Holliday, E.B.; Kocak-Uzel, E.; Feng, L.; Thaker, N.G.; Blanchard, P.; Rosenthal, D.I.; Gunn, G.B.; Garden, A.S.; Frank, S.J. Dosimetric advantages of intensity-modulated proton therapy for oropharyngeal cancer compared with intensity-modulated radiation: A case-matched control analysis. *Med. Dosim.* **2016**, *41*, 189–194. [CrossRef] [PubMed]

76. Sio, T.T.; Lin, H.K.; Shi, Q.; Gunn, G.B.; Cleeland, C.S.; Lee, J.J.; Hernandez, M.; Blanchard, P.; Thaker, N.G.; Phan, J.; et al. Intensity Modulated Proton Therapy Versus Intensity Modulated Photon Radiation Therapy for Oropharyngeal Cancer: First Comparative Results of Patient-Reported Outcomes. *Int. J. Radiat. Oncol. Biol. Phys.* **2016**, *95*, 1107–1114. [CrossRef] [PubMed]

77. Haas-Kogan, D.; Indelicato, D.; Paganetti, H.; Esiashvili, N.; Mahajan, A.; Yock, T.; Flampouri, S.; MacDonald, S.; Fouladi, M.; Stephen, K.; et al. National Cancer Institute Workshop on Proton Therapy for Children: Considerations Regarding Brainstem Injury. *Int. J. Radiat. Oncol. Biol. Phys.* **2018**, *101*, 152–168. [CrossRef] [PubMed]

78. Ebner, D.K.; Kamada, T. The Emerging Role of Carbon-Ion Radiotherapy. *Front. Oncol.* **2016**, *6*, 140. [CrossRef] [PubMed]

79. Mohamad, O.; Imai, R.; Kamada, T.; Nitta, Y.; Araki, N. Carbon ion radiotherapy for inoperable pediatric osteosarcoma. *Oncotarget* **2018**, *9*, 22976–22985. [CrossRef] [PubMed]

80. Blattmann, C.; Oertel, S.; Schulz-Ertner, D.; Rieken, S.; Haufe, S.; Ewerbeck, V.; Unterberg, A.; Karapanagiotou-Schenkel, I.; Combs, S.E.; Nikoghosyan, A.; et al. Non-randomized therapy trial to determine the safety and efficacy of heavy ion radiotherapy in patients with non-resectable osteosarcoma. *BMC Cancer* **2010**, *10*, 96. [CrossRef] [PubMed]

81. Combs, S.E.; Nikoghosyan, A.; Jaekel, O.; Karger, C.P.; Haberer, T.; Munter, M.W.; Huber, P.E.; Debus, J.; Schulz-Ertner, D. Carbon ion radiotherapy for pediatric patients and young adults treated for tumors of the skull base. *Cancer* **2009**, *115*, 1348–1355. [CrossRef] [PubMed]

82. Aggarwal, R.; Yeung, D.; Kumar, P.; Muhlbauer, M.; Kun, L.E. Efficacy and feasibility of stereotactic radiosurgery in the primary management of unfavorable pediatric ependymoma. *Radiother. Oncol. J. Eur. Soc. Ther. Radiol. Oncol.* **1997**, *43*, 269–273. [CrossRef]

83. Hodgson, D.C.; Goumnerova, L.C.; Loeffler, J.S.; Dutton, S.; Black, P.M.; Alexander, E., 3rd; Xu, R.; Kooy, H.; Silver, B.; Tarbell, N.J. Radiosurgery in the management of pediatric brain tumors. *Int. J. Radiat. Oncol. Biol. Phys.* **2001**, *50*, 929–935. [CrossRef]

84. Marcus, K.J.; Goumnerova, L.; Billett, A.L.; Lavally, B.; Scott, R.M.; Bishop, K.; Xu, R.; Young Poussaint, T.; Kieran, M.; Kooy, H.; et al. Stereotactic radiotherapy for localized low-grade gliomas in children: Final results of a prospective trial. *Int. J. Radiat. Oncol. Biol. Phys.* **2005**, *61*, 374–379. [CrossRef] [PubMed]

85. Weintraub, D.; Yen, C.P.; Xu, Z.; Savage, J.; Williams, B.; Sheehan, J. Gamma knife surgery of pediatric gliomas. *J. Neurosurg. Pediatr.* **2012**, *10*, 471–477. [CrossRef] [PubMed]

86. Barcia, J.A.; Barcia-Salorio, J.L.; Ferrer, C.; Ferrer, E.; Algas, R.; Hernandez, G. Stereotactic radiosurgery of deeply seated low grade gliomas. *Acta Neurochir. Suppl.* **1994**, *62*, 58–61. [PubMed]

87. Abe, M.; Tokumaru, S.; Tabuchi, K.; Kida, Y.; Takagi, M.; Imamura, J. Stereotactic radiation therapy with chemotherapy in the management of recurrent medulloblastomas. *Pediatr. Neurosurg.* **2006**, *42*, 81–88. [CrossRef] [PubMed]

88. Patrice, S.J.; Tarbell, N.J.; Goumnerova, L.C.; Shrieve, D.C.; Black, P.M.; Loeffler, J.S. Results of radiosurgery in the management of recurrent and residual medulloblastoma. *Pediatr. Neurosurg.* **1995**, *22*, 197–203. [CrossRef] [PubMed]

89. Barua, K.K.; Ehara, K.; Kohmura, E.; Tamaki, N. Treatment of recurrent craniopharyngiomas. *Kobe J. Med. Sci.* **2003**, *49*, 123–132. [PubMed]

90. Jeon, C.; Kim, S.; Shin, H.J.; Nam, D.H.; Lee, J.I.; Park, K.; Kim, J.H.; Jeon, B.; Kong, D.S. The therapeutic efficacy of fractionated radiotherapy and gamma-knife radiosurgery for craniopharyngiomas. *J. Clin. Neurosci.* **2011**, *18*, 1621–1625. [CrossRef] [PubMed]

91. Niranjan, A.; Kano, H.; Mathieu, D.; Kondziolka, D.; Flickinger, J.C.; Lunsford, L.D. Radiosurgery for craniopharyngioma. *Int. J. Radiat. Oncol. Biol. Phys.* **2010**, *78*, 64–71. [CrossRef] [PubMed]

92. Xu, Z.; Yen, C.P.; Schlesinger, D.; Sheehan, J. Outcomes of Gamma Knife surgery for craniopharyngiomas. *J. Neuro-Oncol.* **2011**, *104*, 305–313. [CrossRef] [PubMed]

93. Foy, A.B.; Wetjen, N.; Pollock, B.E. Stereotactic radiosurgery for pediatric arteriovenous malformations. *Neurosurg. Clin. N. Am.* **2010**, *21*, 457–461. [CrossRef] [PubMed]

94. Hong, J.C.; Salama, J.K. The expanding role of stereotactic body radiation therapy in oligometastatic solid tumors: What do we know and where are we going? *Cancer Treat. Rev.* **2017**, *52*, 22–32. [CrossRef] [PubMed]

95. Kim, H.; Rajagopalan, M.S.; Beriwal, S.; Huq, M.S.; Smith, K.J. Cost-effectiveness analysis of single fraction of stereotactic body radiation therapy compared with single fraction of external beam radiation therapy for palliation of vertebral bone metastases. *Int. J. Radiat. Oncol. Biol. Phys.* **2015**, *91*, 556–563. [CrossRef] [PubMed]

96. Rao, A.D.; Rashid, A.S.; Chen, Q.; Villar, R.C.; Kobyzeva, D.; Nilsson, K.; Dieckmann, K.; Nechesnyuk, A.; Ermoian, R.; Alcorn, S.; et al. Reirradiation for Recurrent Pediatric Central Nervous System Malignancies: A Multi-institutional Review. *Int. J. Radiat. Oncol. Biol. Phys.* **2017**, *99*, 634–641. [CrossRef] [PubMed]

97. Massimino, M.; Miceli, R.; Giangaspero, F.; Boschetti, L.; Modena, P.; Antonelli, M.; Ferroli, P.; Bertin, D.; Pecori, E.; Valentini, L.; et al. Final results of the second prospective AIEOP protocol for pediatric intracranial ependymoma. *Neuro-Oncology* **2016**, *18*, 1451–1460. [CrossRef] [PubMed]

98. Alcorn, S.R.; Chen, M.J.; Claude, L.; Dieckmann, K.; Ermoian, R.P.; Ford, E.C.; Malet, C.; MacDonald, S.M.; Nechesnyuk, A.V.; Nilsson, K.; et al. Practice patterns of photon and proton pediatric image guided radiation treatment: Results from an International Pediatric Research consortium. *Pract. Radiat. Oncol.* **2014**, *4*, 336–341. [CrossRef] [PubMed]

99. Hess, C.B.; Thompson, H.M.; Benedict, S.H.; Seibert, J.A.; Wong, K.; Vaughan, A.T.; Chen, A.M. Exposure Risks Among Children Undergoing Radiation Therapy: Considerations in the Era of Image Guided Radiation Therapy. *Int. J. Radiat. Oncol. Biol. Phys.* **2016**, *94*, 978–992. [CrossRef] [PubMed]

100. Beltran, C.; Krasin, M.J.; Merchant, T.E. Inter- and intrafractional positional uncertainties in pediatric radiotherapy patients with brain and head and neck tumors. *Int. J. Radiat. Oncol. Biol. Phys.* **2011**, *79*, 1266–1274. [CrossRef] [PubMed]

101. Murphy, M.J.; Balter, J.; Balter, S.; BenComo, J.A., Jr.; Das, I.J.; Jiang, S.B.; Ma, C.M.; Olivera, G.H.; Rodebaugh, R.F.; Ruchala, K.J.; et al. The management of imaging dose during image-guided radiotherapy: Report of the AAPM Task Group 75. *Med. Phys.* **2007**, *34*, 4041–4063. [CrossRef] [PubMed]

102. Kornguth, D.; Mahajan, A.; Frija, E.; Chang, E.; Pelloski, C.; Woo, S. 2091: Shape Variability of Craniopharyngioma as Measured on CT-on-Rails During Radiotherapy Treatment. *Int. J. Radiat. Oncol. Biol. Phys.* **2006**, *66*, S259–S260. [CrossRef]

103. Winkfield, K.M.; Linsenmeier, C.; Yock, T.I.; Grant, P.E.; Yeap, B.Y.; Butler, W.E.; Tarbell, N.J. Surveillance of Craniopharyngioma Cyst Growth in Children Treated With Proton Radiotherapy. *Int. J. Radiat. Oncol. Biol. Phys.* **2009**, *73*, 716–721. [CrossRef] [PubMed]

104. Beltran, C.; Naik, M.; Merchant, T.E. Dosimetric effect of target expansion and setup uncertainty during radiation therapy in pediatric craniopharyngioma. *Radiother. Oncol. J. Eur. Soc. Ther. Radiol. Oncol.* **2010**, *97*, 399–403. [CrossRef] [PubMed]

105. Fukunaga-Johnson, N.; Lee, J.H.; Sandler, H.M.; Robertson, P.; McNeil, E.; Goldwein, J.W. Patterns of failure following treatment for medulloblastoma: Is it necessary to treat the entire posterior fossa? *Int. J. Radiat. Oncol. Biol. Phys.* **1998**, *42*, 143–146. [CrossRef]

106. Wolden, S.L.; Dunkel, I.J.; Souweidane, M.M.; Happersett, L.; Khakoo, Y.; Schupak, K.; Lyden, D.; Leibel, S.A. Patterns of failure using a conformal radiation therapy tumor bed boost for medulloblastoma. *J. Clin. Oncol.* **2003**, *21*, 3079–3083. [CrossRef] [PubMed]

107. Michalski, J.M.; Janss, A.; Vezina, G.; Gajjar, A.; Pollack, I.; Merchant, T.E.; FitzGerald, T.J.; Booth, T.; Tarbell, N.J.; Li, Y.; et al. Results of COG ACNS0331: A Phase III Trial of Involved-Field Radiotherapy (IFRT) and Low Dose Craniospinal Irradiation (LD-CSI) with Chemotherapy in Average-Risk Medulloblastoma: A Report from the Children's Oncology Group. *Int. J. Radiat. Oncol. Biol. Phys.* **2016**, *96*, 937–938. [CrossRef]

108. Alapetite, C.; Brisse, H.; Patte, C.; Raquin, M.A.; Gaboriaud, G.; Carrie, C.; Habrand, J.L.; Thiesse, P.; Cuilliere, J.C.; Bernier, V.; et al. Pattern of relapse and outcome of non-metastatic germinoma patients treated with chemotherapy and limited field radiation: The SFOP experience. *Neuro-Oncology* **2010**, *12*, 1318–1325. [PubMed]

109. Calaminus, G.; Kortmann, R.; Worch, J.; Nicholson, J.C.; Alapetite, C.; Garre, M.L.; Patte, C.; Ricardi, U.; Saran, F.; Frappaz, D. SIOP CNS GCT 96: Final report of outcome of a prospective, multinational nonrandomized trial for children and adults with intracranial germinoma, comparing craniospinal irradiation alone with chemotherapy followed by focal primary site irradiation for patients with localized disease. *Neuro-Oncology* **2013**, *15*, 788–796. [PubMed]

110. Rogers, S.J.; Mosleh-Shirazi, M.A.; Saran, F.H. Radiotherapy of localised intracranial germinoma: Time to sever historical ties? *Lancet Oncol.* **2005**, *6*, 509–519. [CrossRef]

111. Khatua, S.; Fangusaro, J.; Dhall, G.; Boyett, J.; Wu, S.; Bartels, U. GC-17THE CHILDREN'S ONCOLOGY GROUP (COG) CURRENT TREATMENT APPROACH FOR CHILDREN WITH NEWLY DIAGNOSED CENTRAL NERVOUS SYSTEM (CNS) LOCALIZED GERMINOMA (ACNS1123 STRATUM 2). *Neuro-Oncology* **2016**, *18*, iii45–iii46. [CrossRef]

112. Goldman, S.; Bouffet, E.; Fisher, P.G.; Allen, J.C.; Robertson, P.L.; Chuba, P.J.; Donahue, B.; Kretschmar, C.S.; Zhou, T.; Buxton, A.B.; et al. Phase II Trial Assessing the Ability of Neoadjuvant Chemotherapy With or Without Second-Look Surgery to Eliminate Measurable Disease for Nongerminomatous Germ Cell Tumors: A Children's Oncology Group Study. *J. Clin. Oncol.* **2015**, *33*, 2464–2471. [CrossRef] [PubMed]

113. Cahlon, O.; Dunkel, I.; Gilheeney, S.; Khakoo, Y.; Souweidane, M.; De Braganca, K.; Kramer, K.; Wolden, S. Craniospinal Radiation Therapy May Not Be Necessary for Localized Nongerminomatous Germ Cell Tumors (NGGCT). *Int. J. Radiat. Oncol. Biol. Phys.* **2014**, *90*, S723–S724. [CrossRef]

114. Packer, R.J.; Goldwein, J.; Nicholson, H.S.; Vezina, L.G.; Allen, J.C.; Ris, M.D.; Muraszko, K.; Rorke, L.B.; Wara, W.M.; Cohen, B.H.; et al. Treatment of children with medulloblastomas with reduced-dose craniospinal radiation therapy and adjuvant chemotherapy: A Children's Cancer Group Study. *J. Clin. Oncol.* **1999**, *17*, 2127–2136. [CrossRef] [PubMed]

115. Goldwein, J.W.; Radcliffe, J.; Johnson, J.; Moshang, T.; Packer, R.J.; Sutton, L.N.; Rorke, L.B.; D'Angio, G.J. Updated results of a pilot study of low dose craniospinal irradiation plus chemotherapy for children under five with cerebellar primitive neuroectodermal tumors (medulloblastoma). *Int. J. Radiat. Oncol. Biol. Phys.* **1996**, *34*, 899–904. [CrossRef]

116. Ramaswamy, V.; Remke, M.; Bouffet, E.; Bailey, S.; Clifford, S.C.; Doz, F.; Kool, M.; Dufour, C.; Vassal, G.; Milde, T.; et al. Risk stratification of childhood medulloblastoma in the molecular era: The current consensus. *Acta Neuropathol.* **2016**, *131*, 821–831. [CrossRef] [PubMed]

© 2018 by the authors. Licensee MDPI, Basel, Switzerland. This article is an open access article distributed under the terms and conditions of the Creative Commons Attribution (CC BY) license (http://creativecommons.org/licenses/by/4.0/).

bioengineering

MDPI

Review

Cutting Edge Therapeutic Insights Derived from Molecular Biology of Pediatric High-Grade Glioma and Diffuse Intrinsic Pontine Glioma (DIPG)

Cavan P. Bailey [1,2,3], Mary Figueroa [1,2,3], Sana Mohiuddin [4], Wafik Zaky [4] and Joya Chandra [1,2,3,*]

1   Department of Pediatrics—Research, University of Texas MD Anderson Cancer Center,
    Houston, TX 77030, USA; cpbailey@mdanderson.org (C.P.B.); mfigueroa1@mdanderson.org (M.F.)
2   The University of Texas MD Anderson Cancer Center, UTHealth Graduate School of Biomedical Sciences,
    Houston, TX 77030, USA
3   Center for Cancer Epigenetics, University of Texas MD Anderson Cancer Center, Houston, TX 77030, USA
4   Department of Pediatrics—Patient Care, University of Texas MD Anderson Cancer Center,
    Houston, TX 77030, USA; smohiuddin@mdanderson.org (S.M.); wzaky@mdanderson.org (W.Z.)
*   Correspondence: jchandra@mdanderson.org; Tel.: +1-713-563-5405

Received: 8 August 2018; Accepted: 15 October 2018; Published: 18 October 2018

Abstract: Pediatric high-grade glioma (pHGG) and brainstem gliomas are some of the most challenging cancers to treat in children, with no effective therapies and 5-year survival at ~2% for diffuse intrinsic pontine glioma (DIPG) patients. The standard of care for pHGG as a whole remains surgery and radiation combined with chemotherapy, while radiation alone is standard treatment for DIPG. Unfortunately, these therapies lack specificity for malignant glioma cells and have few to no reliable biomarkers of efficacy. Recent discoveries have revealed that epigenetic disruption by highly conserved mutations in DNA-packaging histone proteins in pHGG, especially DIPG, contribute to the aggressive nature of these cancers. In this review we pose unanswered questions and address unexplored mechanisms in pre-clinical models and clinical trial data from pHGG patients. Particular focus will be paid towards therapeutics targeting chromatin modifiers and other epigenetic vulnerabilities that can be exploited for pHGG therapy. Further delineation of rational therapeutic combinations has strong potential to drive development of safe and efficacious treatments for pHGG patients.

Keywords: pediatric; high-grade glioma; DIPG; therapeutics; epigenetics; clinical trial

---

1. Pathology of Pediatric High-Grade Gliomas and Diffuse Intrinsic Pontine Gliomas (DIPG)

Malignancies of the central nervous system (CNS) have been a constant challenge for clinicians and researchers alike. Due to their complex morphology and relative resistance to therapies, patient mortality is among the highest of all cancers. Pediatric high-grade gliomas (pHGG) constitute about 11% of all pediatric CNS tumors, with an incidence rate of 0.59 per 100,000 [1]. A subset of pHGG are brainstem gliomas, of which the majority are categorized historically as diffuse intrinsic pontine glioma (DIPG) or diffuse midline glioma (DMG), per the new nomenclature. Five-year survival rates for pHGG as a whole are ~28% [1], but are much more dismal for DIPG at ~2% [2]. HGG typically arise from astrocytic origins, including glial, oligodendrocytes, and ependymal cells. These tumors are classified by the World Health Organization (WHO) as either grade III or IV meaning that they are highly malignant tumors with characteristic findings such as hypercellularity, nuclear atypia, and high mitotic activity with or without microvascular proliferation and pseudopalisading necrosis [3]. HGG include a variety of heterogeneous lesions with differing histology, including anaplastic astrocytoma (WHO Grade III), glioblastoma (WHO grade IV) and diffuse midline glioma, H3-K27M mutant (grade

IV) as the most common types. Other less common types include anaplastic oligodendroglioma (grade III), anaplastic ganglioglioma (grade III), anaplastic pleomorphic xanthoastrocytoma (grade III), giant cell glioblastoma (grade IV), and gliosarcoma (grade IV) [3–5].

2. Molecular Alterations of pHGG and DIPG

Efforts to amass tissue specimens from pHGG patients and apply DNA and RNA sequencing technology have reclassified pHGG pathology. Previously, molecular data was not used to classify brain tumors [3], and for pHGG the only parameters known to correlate with patient outcome are extent of surgical resection and histological grade [6,7]. While useful, these measurements cannot inform pre-clinical molecular research or influence targeted therapeutic decisions. The 2016 WHO reclassification of brain tumors [3] incorporates sequencing data for a more complete picture of pHGG pathology. Broadly, pHGG possess mutations in key epigenetic pathways and signaling kinases, and this new understanding of molecular mechanisms has led to clinical trials designed to target these specific vulnerabilities (Table 1).

Early attempts at large-scale profiling of pediatric gliomas utilized comparative genomic hybridization (CGH) techniques that enable detection of deletions and amplifications in a sample with a resolution of approximately 50 kilobases [8]. This study helped confirm that pediatric high-grade gliomas rely less on *EGFR* amplification, which adult grade IV gliomas often possess, and instead rely more on platelet-derived growth factor receptor (*PDGFR*) amplifications. Another new observation was pediatric high-grade gliomas featuring fewer aberrations in the RTK/PI3K/p53/RB "core signaling pathways" seen in adult glioma. A compelling finding was that a distinct phenotypic category (~20% of their cohort) was found to have a "stable genome" with no detectable copy number alterations. This group not only was in stark contrast to the consistently highly-reconfigured genomes of adult gliomas, but featured improved survival over the "amplified" subtypes in the pediatric range [8]. Presence of disease without extensive genomic rearrangement put forth the idea that perhaps very specific and very powerful mutations could potentially be drivers of gliomagenesis and glioma progression.

Two landmark studies were concurrently published in early 2012 that revealed unique features of pediatric gliomas: they possess mutations in the histone H3 alleles H3.1 (*HIST1H3B*) and H3.3 (*H3F3A*), occurring either at (lysine 27; K to M; H3.1 and H3.3) or near (glycine 34; G to R/V; H3.3 only), a post-translational histone modification point. The K27M variant was found in younger patients (median age, 9 years), midline location including thalamus and brainstem while the G34R/V was more indicative of an adolescent population (median age, 20 years) and hemispheric in location. Due to the co-occurrence of the H3.3 mutation with *TP53* mutations (100% of G34R/V, 82% of K27M) and its conserved nature across patients, this led the authors to believe this is a driver mutation in pHGG [9,10]. These mutations were further found at an even higher prevalence in DIPG, with two independent studies reporting 71% [10] and 78% [11] occurrence of the K27M mutation in a histone H3 variant. Notably, the G34R/V mutation was not found in DIPG, confirming results that it is only seen in supratentorial pHGG. Later findings in 2014 by four separate groups revealed mutations in the activin receptor-like kinase-2 (*ACVR1*) to co-segregate with younger H3.1-K27M, *TP53* wild-type patients whom had slightly improved survival versus other DIPG genotypes [12–15]. All seven studies mentioned the need for further functional validations of these mutations, and encouraged extensive experimental investigation of their mechanistic impact and role in glioma development.

Table 1. Subtypes of pediatric high-grade gliomas, relevant mutations and features, and pediatric clinical trials designed to target these mutations. Diffuse intrinsic pontine glioma (DIPG), histone H3 K27M variant (H3-K27M), TP53, ACVR1, activin receptor-like kinase-2 (ALK2), ATRX-DAXX, platelet-derived growth factor receptor A (PDGFRA), FGFR1, MAPK, BRAF, MEK, NTRK, TRK, mTOR, IDH1, MYCN, TERT, SETD2, WEE1, NCT.

Anatomical Classification	Pediatric High-Grade Gliomas: Mutations, Features, and Novel Clinical Therapies			
	Defining Mutations	Features	Other Mutations	Targeted Therapies and Pediatric Clinical Trials

Midline location	• H3-K27M	• Includes DIPG
		• Predominantly astrocytic differentiation
			• TP53 (60%)
			• ACVR1/ALK2 (20–30%)
			• ATRX-DAXX (30%)
			• PDGFRA amplification
				• H3.3-K27M peptide vaccine
				  ○ NCT02960230 *
				• HDAC inhibitors (Vorinostat, Panobinostat)
				  ○ NCT02420613
				  ○ NCT01189266
				  ○ NCT02717455
				  ○ NCT03566199
				• PDGFR inhibitors (Crenolanib, Dasatinib)
				  ○ NCT01393912
				  ○ NCT01644773
				  ○ NCT00996723

	• FGFR1	• MAPK activation • Usually found in thalamic gliomas		• FGFR inhibitor (Erdafitinib) ○ NCT03210714

	• BRAF V600E	• Rarely co-exists with H3-K27M
		• More common in low grade glioma
				• BRAF/MEK inhibitors (Dabrafinib, Vemurafanib, Trametinib, Selumetinib, Binimetinib)
				  ○ NCT02684058
				  ○ NCT01748149
				  ○ NCT03220035
				  ○ NCT03363217
				  ○ NCT01677741
				  ○ NCT01089101
				  ○ NCT03213691
				  ○ NCT02285439

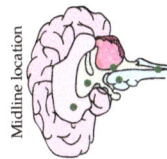

Table 1. *Cont.*

Anatomical Classification	Defining Mutations	Pediatric High-Grade Gliomas: Mutations, Features, and Novel Clinical Therapies		
		Features	Other Mutations	Targeted Therapies and Pediatric Clinical Trials
	• H3-G34R/V	• 15% • Histologically homogenous appearance	• ATRX/DAXX (100%) • TP53 (90%) • NTRK fusion	• TRK inhibitors (Larotectinib, Entrectinib) ○ NCT03213704 ○ NCT02637687 ○ NCT03215511 ○ NCT02650401
Hemispheric location	• IDH1 mutant	• Small population • Better survival than wild type		• mTOR inhibitor (Everolimus) ○ NCT01734512
	• H3/IDH wild type		• MYCN amplification • TP53 • PDGFRA • TERT	
	• SETD2 (15%)	• Mutually exclusive with H3-G34R/V		• WEE1 inhibitor (Adavosertib)

* All NCT entries indicate trial number listed in clinicaltrials.gov.

Other mutations co-occur with the histone mutations in pHGG, including *ATRX*, *SETD2*, and *TP53* mutations (Table 1). Adult HGG, which do not possess histone mutations, are more likely to have *CDKN2A/B* deletions, *IDH1/2* mutations, *EGFR* amplification and *PTEN* loss. DIPGs follow the pHGG molecular pattern but are further enriched for *TP53* mutation and *PTEN* loss, lack *SETD2* mutations, and possess *PDGFRA* amplification (Table 1) [16,17]. Alterations in cell growth pathways are seen in some pHGGs, including *BRAF* and *FGFR1* mutations and gene fusions among the *TRK* family (*NTRK1/2/3*) (Table 1) [18]; however, *BRAF/FGFR1* perturbations are far more prominent in pediatric low-grade gliomas (pLGGs) [19].

3. Histone H3-K27M and G34R/V-Specific Mechanisms in pHGG

A rigorous examination of histone H3 amino acid substitutions using in vitro biochemical assays, HEK-293 cell culture, and murine brain stem glioma models found that the K27M mutation can inactivate the polycomb repressive complex 2 (PRC2) via methionine/isoleucine-specific hydrophobic interactions with the active site of enhancer of zeste homolog 2 (EZH2) [20]. As such, the positive feedback loop regulating PRC2 activity is disrupted and global hypomethylation at H3K27 is observed, along with increased H3K27 acetylation. It was later found that H3.1/3.3-K27M inactivation of PRC2 can also enrich H3K27me3 at specific loci, causing either gene activation (hypo-H3K27me3) or gene suppression (enriched-H3K27me3) in unique pathways [21]. These findings were recapitulated with additional insight into the role of DNA methylation as a consequence of H3.3-K27M, as well as observance of increased H3K27me3 in intergenic regions of the genome, suggesting a possible link with miRNA or lncRNA dysregulation in H3-mutated pHGG [22].

While the K27M mutations possess a potent inhibitory effect on SET-domain containing methyltransferases, the G34R/V mutation did not recapitulate the same effect in vivo at H3K36. However, immunoprecipitated nucleosomes from the same cells showed lowered H3K36me2/3 on the G34R/V-H3.2/3 histone, but not the wild-type H3.2/3 histone in the same oligonucleosome chain [20]. SETD2 is the methyltransferase responsible for generating the H3K36me3 mark, and its interaction with G34R/V mutations has not been extensively studied (Figure 1). The H3.3K36me3 reader protein ZMYND11 cannot bind under G34R/V conditions [23]; however, further investigations of protein complex, reader, and SETD2 binding with G34R/V nucleosomes are needed to provide insight into why the G34R/V mutation does not create a hypomethylated phenotype at the H3K36 residue.

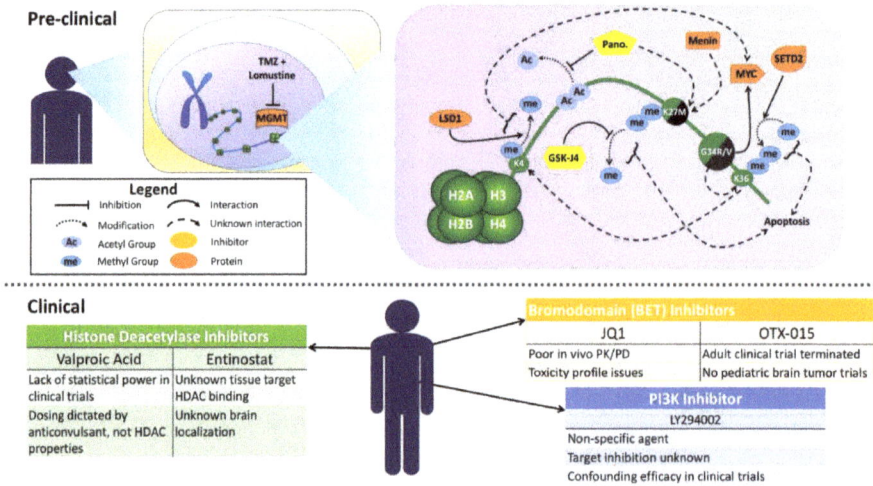

Figure 1. Unexplored pre-clinical molecular interactions and inconclusive clinical data in pediatric high-grade glioma research and trials. Temozolomide (TMZ), O-6-methylguanine-DNA methyltransferase (MGMT), lysine-specific demethylase 1 (LSD1), Pano, MYC, GSK-J4, JQ1, PK/PD, OTX-015, PI3K.

4. Standard of Care Therapies in pHGG including DIPG

The outcome for children with HGG remains dismal despite the use of multi-modal therapy including surgery, radiation therapy (RT) and chemotherapy. In a phase II trial that used temozolomide (TMZ) in combination with lomustine following RT, the authors observed an increase in the overall survival (OS) and event-free survival (EFS) when compared to TMZ + RT alone, but this came with higher toxicities [24]. They hypothesize treatment with multiple alkylating agents can degrade expression of de-alkylating enzymes and increase sensitivity when a second agent is used, but this has not been verified experimentally (Figure 1). Anti-angiogenic therapy with bevacizumab has shown efficacy in recurrent adult glioma [25], but not in newly diagnosed adult glioma patients [26,27] or in combination with lomustine [28]. In pHGG patients, bevacizumab was evaluated in addition to TMZ + RT in patients with newly diagnosed non-brainstem high-grade glioma in a phase II open-label, randomized, multicenter trial (HERBY) [29]. A follow-up molecular analysis of the HERBY trial found addition of bevacizumab to TMZ + RT conferred a survival benefit only in hypermutated and *BRAF*-driven tumors that display an increase in CD8+ T-cell infiltration after therapy [30]. Notably, K27M and G34R/V tumors were "immunologically cold" in their analysis, highlighting how checkpoint blockade and other T-cell-stimulating therapies may not be a viable option for these patients.

DIPG presents a special challenge as their diffuse nature in the brainstem makes surgery not feasible [31,32]. Recent studies have shown MRI-guided stereotactic biopsy to be a safe and reliable method of molecularly characterizing a DIPG patient [33,34], and others were able to detect mutational status using cerebrospinal fluid-sourced tumor DNA [35]. Radiotherapy for DIPG can prolong survival by a few months, but ultimately RT is palliative and patients succumb to progressive disease within a year. Multiple lines of treatment have been tested in clinical trials including chemotherapy and targeted therapy without improving outcome to standard radiation alone [36]. From a clinical therapeutic perspective, it is difficult to discriminate whether the resistance of DIPG to chemotherapy is due to inherent cellular mechanisms or poor delivery of the drug to the tumor site (Table 1). This issue will be addressed later in the review by exploring cutting-edge modalities to improve drug delivery to the brain.

Studies initially focused on TMZ resistance in pHGG led researchers to the homeobox genes *HOXA9* and *HOXA10*, which were found to be upregulated in pediatric cell lines and correlated to TMZ resistance independent of the demethylating gene; O-6-methylguanine-DNA methyltransferase (*MGMT*) [37]. The pediatric TMZ-resistant cells were also found to be high in progenitor cell markers such as Nestin and CD133, possibly endowing these cells with enhancements in self-renewal and other stem-like properties. Further investigation of the *HOX* signature in adult glioma found that DNA methylation combined with chromosome 7 copy gains endow the adult glioma cells with aberrant *HOX* expression which is normally never seen past the hindbrain [38].

Similar dysregulation of *HOX* expression was also found in adult high-grade astrocytoma and here *HOXA9/10* could be regulated in a PI3K-dependent manner that relied on epigenetic modifications. In their analysis, usage of a PI3K inhibitor decreased *HOX* expression while also increasing trimethylation of lysine 27 on histone H3 (H3K27me3) and decreasing trimethylation of lysine 4 on histone H3 (H3K4me3) [39]. It should be noted the authors used LY294002 to inhibit PI3K, a well-documented non-specific agent [40] that also inhibits CK2, mTOR, and GSK3β [41] along with bromodomains [42] (Figure 1). More recently, targeting of the PI3K pathway in DIPG either with direct chemical agents [43, 44] or by modulating the tumor microenvironment [45], has demonstrated efficacy in pre-clinical studies that have not been replicated in a small clinical trial [46]. However, there exists a wide range of drugs designed to target this pathway, and the authors note they did not measure target inhibition in their patient cohort [46].

5. Histone H3-K27M and G34R/V-Specific Therapies

With the recent mechanistic insights into how mutant histones poise the epigenome for later tumor development [47], thoughts turned to how to therapeutically target these phenomena and potentially bring new hope to patients with untreatable disease. The well-known oncogene *MYC* was one of the first targets, identified because the G34R/V mutation in H3.3 leads to enrichment of the activating mark H3K36me3 across the genome, with MYC's genomic locus being the most differentially affected v. H3.3 wild-type [48]. A synthetic lethality screen of 714 kinases in pediatric gliomas identified checkpoint kinase 1 (CHK1) and aurora kinase A (AURKA), both MYC stabilizing proteins, and usage of a small molecule inhibitor of AURKA (VX-689) exhibited potent cell killing in vitro (Table 2). Early neural development genes were also shown to be G34R/V activated, and a later study revealed MYC alone could induce stem-like phenotypes dependent on H3K4 methylation levels regulated by lysine-specific demethylase 1 (LSD1) [49]. Whether the G34R/V mechanism of tumorigenesis is dependent on, concurrent with, or completely independent of the LSD1-MYC interaction remains to be investigated (Figure 1).

Table 2. Experimental pre-clinical therapeutics for pediatric high-grade gliomas including compound name, target of therapy, rationale for use in this disease, and references.

Pre-Clinical Therapies for Pediatric High-Grade Gliomas			
Compound(s)	Target	Rationale	Reference
VX-689 (renamed MK-5108)	AURKA (Aurora kinase A)	Destabilizes MYC	[48]
MI-2	MEN1 (Menin)	Blocks menin-MLL-AF9 initiated leukemic oncogenesis; exact role in glioma undefined	[50,51]
GSK-J4	JMJD3 (Jumonji-domain demethylase)	Prevents further demethylation of H3K27 mark in H3-K27M mutated glioma	[52]
OTX-015 JQ1	BRD2/3/4 (Bromodomain-containing proteins)	Interrupts BRD → H3-acetylation binding that is increased by H3.3-K27M	[53–55]
GSK126 GSK343 EPZ-6438	EZH2 (Enhancer of zeste homolog 2)	De-represses tumor suppressor $p16^{INK4a}$ and induces apoptosis	[55–57]
BMS-754807	Multi-kinase, most potent against IGF-1R (Insulin-like growth factor 1 receptor)	Effective in compound screen; multiple valid targets in DIPG	[58]
PTC-209	BMI-1 (polycomb group RING finger protein 4)	Induces cell cycle arrest and telomerase downregulation, reduces migration, increases sensitivity to radiotherapy	[59,60]

Given the role of G34R/V in modulating early neural oncogenesis, it remained to be seen if the K27M mutation would result in a similar fate. Using human embryonic stem cells (hESCs) as a model system, it was observed that transduced H3.3-K27M in combination with p53 silencing and constitutively active platelet-derived growth factor receptor A (PDGFRA) conferred neoplastic traits to the cells. A small-molecule screen of >80 epigenetic compounds revealed menin (MEN1) as a potent target specific to K27M-possessing cells [50]. Menin is found in the trithorax histone methyltransferase complex and functions as either a tumor suppressor or an oncogene depending on the cell type. Further study is needed to determine the mechanistic role it plays in H3.3-K27M pediatric glioma as well as a safety/efficacy profile for the MEN1 inhibitor MI-2 [51] (Figure 1, Table 2). Screens on the genetic level using CRISPR-Cas9 have also been used in adult glioma models [61], setting the stage for H3.1/3.3-specific investigation and potential gene therapy applications.

Potential also exists to target epigenetic proteins that participate in the pathways interrupted by H3-K27M-mediated PRC2 inactivation. The hypomethylation signature at H3K27 can potentially be reversed by inhibiting histone H3 demethylases, in particular the ubiquitously transcribed tetratricopeptide repeat, X chromosome (UTX) and Jumonji-domain 3 (JMJD3) demethylases, which are specific to the H3K27 mark. A small-molecule inhibitor of UTX/JMJD3, dubbed GSKJ4, was found to have efficacy in cell lines expressing H3.3-K27M, but displayed no effect in H3.3-G34R/V or wild-type H3.3 controls [52] (Table 2). The exact mechanistic action of the growth inhibition and apoptosis induced by GSKJ4 remains to be explored, as changes in H3K27me2/3 status can broadly affect gene expression (Figure 1). While PRC2 inactivation has been a prime molecular focus of DIPG mutations, the PRC1 complex has recently come to light as a therapeutic target. Single-cell RNA-Seq studies defined the PRC1 subunit BMI-1 to be upregulated in K27M DIPG cancer stem cells with an oligodendrocyte precursor (OPC) phenotype [60]. The authors and others [59] could target BMI-1 with the small molecule PTC-209 and found strong combinatorial effects with *PDGFRA* CRISPR-Cas9 knockouts (Table 2). As such, the more clinically advanced BMI-1 inhibitor PTC-596 is currently in clinical trials for DIPG and pHGG (NCT03605550).

Chromatin-association studies using *Drosophila melanogaster* confirmed the results of Lewis et al. [20], finding H3.3-K27M-containing nucleosomes to be enriched in H3K27 acetylation. A key additional observation was increased presence on these same nucleosomes of bromodomain and extra-terminal motif (BET) proteins BRD1 and BRD4 which function to "read" histone acetylation marks [62]. This suggests use of inhibitors of the BRD family as a therapeutic option in K27M-mutated DIPG, and indeed the pan-BET inhibitor JQ1 has displayed anti-tumor efficacy in pre-clinical models [53,54] (Table 2). Concurrent research examined the use of EZH2 inhibitors, either as a single agent [56] or in combination with BET inhibitors [55]. The mechanism of EZH2 inhibitor tumor suppression was found to be re-expression of p16 and subsequent apoptosis [56], but this could not be recapitulated in a genetically-engineered mouse model of H3.3-K27M DIPG [57] (Table 2). Of note is that the oncogenic driver mutation *BRAF*-V600E, present in a small subset of pHGG, may also benefit from epigenetic therapy. Recent studies have shown *EZH2* to be an oncogene in *BRAF*-driven melanoma [63], and combination therapy of the BRAF inhibitor vemurafinib with JQ1 [64] or DNA methyltransferase inhibitor decitabine [65] displayed efficacy in pre-clinical melanoma models. In clinical studies, a Phase IIa clinical trial of the pan-BET inhibitor OTX-015 in recurrent adult glioblastoma (NCT02296476) was terminated prior to completion, and a pediatric screening trial that includes an EZH2 inhibitor is currently recruiting (NCT03155620). As such, the potential of BET and EZH2 inhibition in pHGG populations remains inconclusive (Figure 1).

High-throughput approaches to find viable "hits" further validated the use of demethylase inhibitors, with GSKJ4 displaying a strong synergistic effect when combined with the pan-histone deacetylase inhibitor (HDACi) panobinostat [66]. As expected, use of these drugs together increased H3 acetylation along with H3K27 methylation. Biochemical analyses of K27M-PRC2 interactions demonstrated that acetylation of H3 at various lysine residues could reverse PRC2 inactivation by over 80-fold [67], but whether this is the main mechanism of panobinostat in K27M models remains to be determined. HDAC inhibitors have previously been shown to be effective against adult glioma [68] and glial stem cells [69], with their efficacy enhanced by inhibiting LSD1. It is currently unknown if dual inhibition of LSD1 (H3K4 demethylation) and UTX/JMJD3 (H3K27 demethylation) can influence HDAC inhibitor-induced cell death, either in adult glioma or H3-mutated pediatric DIPG.

While Grasso et al. [66] used human xenografts of DIPG, later use of panobinostat in an immune-competent genetically-engineered mouse model of DIPG replicated the in vitro potency of panobinostat but found dose-limiting toxicities when administered in vivo. Panobinostat is well known to cause peripheral thrombocytopenia [70,71], which is not unexpected as it was initially FDA-approved for the blood cancer multiple myeloma. Recent clinical trial data presented at ISPNO 2018 confirms the dose-limiting toxicities of panobinostat when administered peripherally in pediatric patients [72]. Other HDAC inhibitors have been tested in pHGG; however, a retrospective study of valproic acid found no significant enhancement of treatments including radiation and DNA damaging and demethylating agents [73]. Lack of statistical power and dosage of valproic acid as an anticonvulsant, instead of for its HDAC properties, leaves the use of valproic acid in pHGG inconclusive on the clinical end (Figure 1). Given that the only positive in vivo data using HDAC inhibitors in pHGG utilized convection-enhanced delivery directly to the pons [66], alternative delivery strategies of small molecules and biologics to the dense tumor tissue may hold the key to successful pHGG therapy.

6. Therapeutic Delivery

A crucial challenge in glioma drug development is sufficient delivery to the brain, involving passage through the blood-brain barrier (BBB) when a compound is given systemically. Recent work has shown brainstem gliomas have less than one-third the tumor volume and possess lower rates of BBB permeability versus cortical gliomas, independent of H3.3 mutation status [74]. Panobinostat and GSKJ4 were verified in murine brain tissue via paired high-performance liquid chromatography (HPLC)/mass spectrometry (MS) [52,66] but as of yet only panobinostat is in pediatric trials (Table 1). The pharmacokinetics of VX-689, MI-2, OTX-015, EZH2 inhibitors, and PTC-596 in the brain remain unknown, while studies on a multi-kinase drug (BMS-754807) demonstrated efficacy against

H3.3-K27M murine cells in vitro but failed to reach IC_{50} levels in tumor tissue when dosed at 50 mg/kg in vivo [58] (Table 2).

Given these limitations, researchers need to think outside the current small-molecule therapeutic box. One method being explored is focused ultrasound (FUS), which aims to temporarily disrupt the tight junctions in the BBB to allow passage of drugs to the tumor site. It has shown limited efficacy in enhancement of temozolomide treatment [75], and a clinical trial is being pursued using doxorubicin (NCT02343991). This method can also be combined with "microbubbles" to further open the BBB [76], but FUS has not been tested with the epigenetic agents listed above. An even more exciting advancement uses a chain of iron-oxide nanospheres attached to a doxorubicin-loaded liposome. The nanospheres are coated with cyclic arginylglycylaspartic acid (RGD) and targeted to the $\alpha_v\beta_3$ integrin peptide that is overexpressed in brain tumor vasculature. This flexible "tail" of three iron-oxide spheres increases binding to the endothelial cells, and drug is subsequently released by a localized radio frequency (RF) field that the vibrates the tail and breaks open the liposome [77]. Further studies could adapt this technology to carry other drugs relevant to H3-mutated pediatric tumors, as well as increase drug penetrance to difficult-to-reach areas of the brain such as the pons in DIPG patients.

7. Future Directions

While we have explored promising small molecule approaches to treat pHGG, we must consider what role the immuno-oncology revolution will play in pediatric glioma patients going forward. The brain was previously thought to be an immune-privileged site, but this has been proven to not be the case [78]. All cells generate antigens on their cell surface corresponding to internal proteins, and recently researchers were able to isolate K27M-specific T-cells from DIPG patients [79], proving that the immune system can "see" the tumor and generate a reaction specific to its mutational status. Another K27M-driven immunotherapy target in DIPG is the lipid surface marker GD2, for which use of chimeric-antigen receptor T-cells (CAR-T) shows promise in pre-clinical models, but also displays cautionary inflammatory toxicities [80,81]. Furthermore, the potential use of expanded panels of epigenetic agents has been understudied in how it would affect immune function, both systemically and in the tumor microenvironment [82]. There also exists a tight interplay between cellular metabolism and epigenetic functions [83]; there may be other possibilities for modulation of these pathways through nutrient control and metabolic pathway inhibitors.

A large gap in knowledge of the biology of pHGG is in whether the epigenetic and kinase alterations observed in fully-formed tumors contribute to early oncogenesis of the tumor, or if they are only necessary for mature tumor maintenance. Kinase signaling through FGFR1 [84,85] and PDGFR [86] are crucial to normal CNS development as well as being potentially involved in recovery from radiation-induced necrosis [87], while the role of ACVR1 [88] in CNS function remains more mysterious. For H3.1/3.3 mutations, the highly conserved genotype for age and brain location points to proper H3 modification being involved in gene expression regulation during CNS maturation. It is possible that H3K27 methylation/acetylation is more important infratentorially while H3K36 methylation/acetylation is associated with supratentorial regulation. Complex in vivo models could answer these questions by modifying histones or their associated remodelers over periods of cortical maturation.

It is now understood adult and pediatric gliomas are distinct at the molecular level, a fact that was previously invisible pathologically and can be attributed to the genomics era of low-cost, high-quality sequencing. Rigorous and innovative research has shed new light on how epigenetic and kinase mutations contribute to pHGG molecular signaling. Standard of care currently remains unchanged, but rapidly developing targeted therapies and drug delivery methods are moving into clinical trials and may shift the treatment paradigm of pHGG in the coming years. There are several exciting avenues to pursue in pediatric glioma research, and fresh ideas and new approaches are desperately needed to combat this devastating illness.

Author Contributions: Conceptualization, C.P.B. and J.C.; Writing-Original Draft Preparation, C.P.B.; Writing-Review & Editing, S.M., W.Z., and J.C.; Visualization, M.F.; Supervision, J.C.

Funding: This research was funded by NIH/NINDS R21 NS093387, NIH/NCI Brain Cancer SPORE P50 CA127001, the Team Connor Foundation, the Marnie Rose Foundation, and the Thomas Scott Family Foundation, and supported by the NIH/NCI under award number P30CA016672. C.P.B. was funded by a Center for Cancer Epigenetics (CCE) Scholar award from MD Anderson Cancer Center.

Acknowledgments: We thank Richard R. Behringer, Rachel K. Miller, and Eric C. Swindell for helpful comments on the manuscript as well as classmates of C.P.B. for their unwavering support.

Conflicts of Interest: The authors declare no conflict of interest.

References

1. Ostrom, Q.T.; de Blank, P.M.; Kruchko, C.; Petersen, C.M.; Liao, P.; Finlay, J.L.; Stearns, D.S.; Wolff, J.E.; Wolinsky, Y.; Letterio, J.J.; et al. Alex's Lemonade Stand Foundation Infant and Childhood Primary Brain and Central Nervous System Tumors Diagnosed in the United States in 2007–2011. *Neuro Oncol.* **2015**, *16* (Suppl. 10), x1–x36. [CrossRef]
2. Hoffman, L.M.; Veldhuijzen van Zanten, S.E.M.; Colditz, N.; Baugh, J.; Chaney, B.; Hoffmann, M.; Lane, A.; Fuller, C.; Miles, L.; Hawkins, C.; et al. Clinical, Radiologic, Pathologic, and Molecular Characteristics of Long-Term Survivors of Diffuse Intrinsic Pontine Glioma (DIPG): A Collaborative Report From the International and European Society for Pediatric Oncology DIPG Registries. *J. Clin. Oncol.* **2018**, *36*, 1963–1972. [CrossRef] [PubMed]
3. Louis, D.N.; Perry, A.; Reifenberger, G.; von Deimling, A.; Figarella-Branger, D.; Cavenee, W.K.; Ohgaki, H.; Wiestler, O.D.; Kleihues, P.; Ellison, D.W. The 2016 World Health Organization Classification of Tumors of the Central Nervous System: A summary. *Acta Neuropathol.* **2016**, *131*, 803–820. [CrossRef] [PubMed]
4. Merchant, T.E.; Pollack, I.F.; Loeffler, J.S. Brain tumors across the age spectrum: Biology, therapy, and late effects. *Semin. Radiat. Oncol.* **2010**, *20*, 58–66. [CrossRef] [PubMed]
5. Khatua, S.; Zaky, W. Diffuse intrinsic pontine glioma: Time for therapeutic optimism. *CNS Oncol.* **2014**, *3*, 337–348. [CrossRef] [PubMed]
6. Fangusaro, J. Pediatric high grade glioma: A review and update on tumor clinical characteristics and biology. *Front. Oncol.* **2012**, *2*, 105. [CrossRef] [PubMed]
7. Chamdine, O.; Gajjar, A. Molecular characteristics of pediatric high-grade gliomas. *CNS Oncol.* **2014**, *3*, 433–443. [CrossRef] [PubMed]
8. Bax, D.A.; Mackay, A.; Little, S.E.; Carvalho, D.; Viana-Pereira, M.; Tamber, N.; Grigoriadis, A.E.; Ashworth, A.; Reis, R.M.; Ellison, D.W.; et al. A distinct spectrum of copy number aberrations in pediatric high-grade gliomas. *Clin. Cancer Res.* **2010**, *16*, 3368–3377. [CrossRef] [PubMed]
9. Wu, G.; Broniscer, A.; McEachron, T.A.; Lu, C.; Paugh, B.S.; Becksfort, J.; Qu, C.; Ding, L.; Huether, R.; Parker, M.; et al. Somatic histone H3 alterations in pediatric diffuse intrinsic pontine gliomas and non-brainstem glioblastomas. *Nat. Genet.* **2012**, *44*, 251–253. [CrossRef] [PubMed]
10. Schwartzentruber, J.; Korshunov, A.; Liu, X.Y.; Jones, D.T.; Pfaff, E.; Jacob, K.; Sturm, D.; Fontebasso, A.M.; Quang, D.A.; Tonjes, M.; et al. Driver mutations in histone H3.3 and chromatin remodelling genes in paediatric glioblastoma. *Nature* **2012**, *482*, 226–231. [CrossRef] [PubMed]
11. Khuong-Quang, D.A.; Buczkowicz, P.; Rakopoulos, P.; Liu, X.Y.; Fontebasso, A.M.; Bouffet, E.; Bartels, U.; Albrecht, S.; Schwartzentruber, J.; Letourneau, L.; et al. K27M mutation in histone H3.3 defines clinically and biologically distinct subgroups of pediatric diffuse intrinsic pontine gliomas. *Acta Neuropathol.* **2012**, *124*, 439–447. [CrossRef] [PubMed]
12. Taylor, K.R.; Mackay, A.; Truffaux, N.; Butterfield, Y.; Morozova, O.; Philippe, C.; Castel, D.; Grasso, C.S.; Vinci, M.; Carvalho, D.; et al. Recurrent activating ACVR1 mutations in diffuse intrinsic pontine glioma. *Nat. Genet.* **2014**, *46*, 457–461. [CrossRef] [PubMed]
13. Wu, G.; Diaz, A.K.; Paugh, B.S.; Rankin, S.L.; Ju, B.; Li, Y.; Zhu, X.; Qu, C.; Chen, X.; Zhang, J.; et al. The genomic landscape of diffuse intrinsic pontine glioma and pediatric non-brainstem high-grade glioma. *Nat. Genet.* **2014**, *46*, 444–450. [CrossRef] [PubMed]

14. Fontebasso, A.M.; Papillon-Cavanagh, S.; Schwartzentruber, J.; Nikbakht, H.; Gerges, N.; Fiset, P.O.; Bechet, D.; Faury, D.; De Jay, N.; Ramkissoon, L.A.; et al. Recurrent somatic mutations in ACVR1 in pediatric midline high-grade astrocytoma. *Nat. Genet.* **2014**, *46*, 462–466. [CrossRef] [PubMed]

15. Buczkowicz, P.; Hoeman, C.; Rakopoulos, P.; Pajovic, S.; Letourneau, L.; Dzamba, M.; Morrison, A.; Lewis, P.; Bouffet, E.; Bartels, U.; et al. Genomic analysis of diffuse intrinsic pontine gliomas identifies three molecular subgroups and recurrent activating ACVR1 mutations. *Nat. Genet.* **2014**, *46*, 451–456. [CrossRef] [PubMed]

16. Schroeder, K.M.; Hoeman, C.M.; Becher, O.J. Children are not just little adults: Recent advances in understanding of diffuse intrinsic pontine glioma biology. *Pediatr. Res.* **2014**, *75*, 205–209. [CrossRef] [PubMed]

17. Mackay, A.; Burford, A.; Carvalho, D.; Izquierdo, E.; Fazal-Salom, J.; Taylor, K.R.; Bjerke, L.; Clarke, M.; Vinci, M.; Nandhabalan, M.; et al. Integrated Molecular Meta-Analysis of 1,000 Pediatric High-Grade and Diffuse Intrinsic Pontine Glioma. *Cancer Cell* **2017**, *32*. [CrossRef] [PubMed]

18. Vanan, M.I.; Eisenstat, D.D. Management of high-grade gliomas in the pediatric patient: Past, present, and future. *Neurooncol. Pract.* **2014**, *1*, 145–157. [CrossRef] [PubMed]

19. Johnson, A.; Severson, E.; Gay, L.; Vergilio, J.A.; Elvin, J.; Suh, J.; Daniel, S.; Covert, M.; Frampton, G.M.; Hsu, S.; et al. Comprehensive Genomic Profiling of 282 Pediatric Low- and High-Grade Gliomas Reveals Genomic Drivers, Tumor Mutational Burden, and Hypermutation Signatures. *Oncologist* **2017**, *22*, 1478–1490. [CrossRef] [PubMed]

20. Lewis, P.W.; Muller, M.M.; Koletsky, M.S.; Cordero, F.; Lin, S.; Banaszynski, L.A.; Garcia, B.A.; Muir, T.W.; Becher, O.J.; Allis, C.D. Inhibition of PRC2 activity by a gain-of-function H3 mutation found in pediatric glioblastoma. *Science* **2013**, *340*, 857–861. [CrossRef] [PubMed]

21. Chan, K.M.; Fang, D.; Gan, H.; Hashizume, R.; Yu, C.; Schroeder, M.; Gupta, N.; Mueller, S.; James, C.D.; Jenkins, R.; et al. The histone H3.3K27M mutation in pediatric glioma reprograms H3K27 methylation and gene expression. *Genes Dev.* **2013**, *27*, 985–990. [CrossRef] [PubMed]

22. Bender, S.; Tang, Y.; Lindroth, A.M.; Hovestadt, V.; Jones, D.T.; Kool, M.; Zapatka, M.; Northcott, P.A.; Sturm, D.; Wang, W.; et al. Reduced H3K27me3 and DNA hypomethylation are major drivers of gene expression in K27M mutant pediatric high-grade gliomas. *Cancer Cell* **2013**, *24*, 660–672. [CrossRef] [PubMed]

23. Wen, H.; Li, Y.; Xi, Y.; Jiang, S.; Stratton, S.; Peng, D.; Tanaka, K.; Ren, Y.; Xia, Z.; Wu, J.; et al. ZMYND11 links histone H3.3K36me3 to transcription elongation and tumour suppression. *Nature* **2014**, *508*, 263–268. [CrossRef] [PubMed]

24. Jakacki, R.I.; Cohen, K.J.; Buxton, A.; Krailo, M.D.; Burger, P.C.; Rosenblum, M.K.; Brat, D.J.; Hamilton, R.L.; Eckel, S.P.; Zhou, T.; et al. Phase 2 study of concurrent radiotherapy and temozolomide followed by temozolomide and lomustine in the treatment of children with high-grade glioma: A report of the Children's Oncology Group ACNS0423 study. *Neuro Oncol.* **2016**, *18*, 1442–1450. [CrossRef] [PubMed]

25. Friedman, H.S.; Prados, M.D.; Wen, P.Y.; Mikkelsen, T.; Schiff, D.; Abrey, L.E.; Yung, W.K.; Paleologos, N.; Nicholas, M.K.; Jensen, R.; et al. Bevacizumab alone and in combination with irinotecan in recurrent glioblastoma. *J. Clin. Oncol.* **2009**, *27*, 4733–4740. [CrossRef] [PubMed]

26. Chinot, O.L.; Wick, W.; Mason, W.; Henriksson, R.; Saran, F.; Nishikawa, R.; Carpentier, A.F.; Hoang-Xuan, K.; Kavan, P.; Cernea, D.; et al. Bevacizumab plus radiotherapy-temozolomide for newly diagnosed glioblastoma. *N. Engl. J. Med.* **2014**, *370*, 709–722. [CrossRef] [PubMed]

27. Gilbert, M.R.; Dignam, J.J.; Armstrong, T.S.; Wefel, J.S.; Blumenthal, D.T.; Vogelbaum, M.A.; Colman, H.; Chakravarti, A.; Pugh, S.; Won, M.; et al. A randomized trial of bevacizumab for newly diagnosed glioblastoma. *N. Engl. J. Med.* **2014**, *370*, 699–708. [CrossRef] [PubMed]

28. Wick, W.; Gorlia, T.; Bendszus, M.; Taphoorn, M.; Sahm, F.; Harting, I.; Brandes, A.A.; Taal, W.; Domont, J.; Idbaih, A.; et al. Lomustine and Bevacizumab in Progressive Glioblastoma. *N. Engl. J. Med.* **2017**, *377*, 1954–1963. [CrossRef] [PubMed]

29. Grill, J.; Massimino, M.; Bouffet, E.; Azizi, A.A.; McCowage, G.; Cañete, A.; Saran, F.; Le Deley, M.C.; Varlet, P.; Morgan, P.S.; et al. Phase II, Open-Label, Randomized, Multicenter Trial (HERBY) of Bevacizumab in Pediatric Patients With Newly Diagnosed High-Grade Glioma. *J. Clin. Oncol.* **2018**, *36*, 951–958. [CrossRef] [PubMed]

30. Mackay, A.; Burford, A.; Molinari, V.; Jones, D.T.W.; Izquierdo, E.; Brouwer-Visser, J.; Giangaspero, F.; Haberler, C.; Pietsch, T.; Jacques, T.S.; et al. Molecular, Pathological, Radiological, and Immune Profiling of Non-brainstem Pediatric High-Grade Glioma from the HERBY Phase II Randomized Trial. *Cancer Cell* **2018**, *33*, 829–842.e5. [CrossRef] [PubMed]

31. Xu, C.; Liu, X.; Geng, Y.; Bai, Q.; Pan, C.; Sun, Y.; Chen, X.; Yu, H.; Wu, Y.; Zhang, P.; et al. Patient-derived DIPG cells preserve stem-like characteristics and generate orthotopic tumors. *Oncotarget* **2017**, *8*, 76644–76655. [CrossRef] [PubMed]

32. Zhang, L.; Pan, C.-C.; Li, D. The historical change of brainstem glioma diagnosis and treatment: From imaging to molecular pathology and then molecular imaging. *Chin. Neurosurg. J.* **2015**, *1*, 4. [CrossRef]

33. Hamisch, C.; Kickingereder, P.; Fischer, M.; Simon, T.; Ruge, M.I. Update on the diagnostic value and safety of stereotactic biopsy for pediatric brainstem tumors: A systematic review and meta-analysis of 735 cases. *J. Neurosurg. Pediatr.* **2017**, *20*, 261–268. [CrossRef] [PubMed]

34. Carai, A.; Mastronuzzi, A.; De Benedictis, A.; Messina, R.; Cacchione, A.; Miele, E.; Randi, F.; Esposito, G.; Trezza, A.; Colafati, G.S.; et al. Robot-Assisted Stereotactic Biopsy of Diffuse Intrinsic Pontine Glioma: A Single-Center Experience. *World Neurosurg.* **2017**, *101*, 584–588. [CrossRef] [PubMed]

35. Huang, T.Y.; Piunti, A.; Lulla, R.R.; Qi, J.; Horbinski, C.M.; Tomita, T.; James, C.D.; Shilatifard, A.; Saratsis, A.M. Detection of Histone H3 mutations in cerebrospinal fluid-derived tumor DNA from children with diffuse midline glioma. *Acta Neuropathol. Commun.* **2017**, *5*, 28. [CrossRef] [PubMed]

36. Clymer, J.; Kieran, M.W. The Integration of Biology Into the Treatment of Diffuse Intrinsic Pontine Glioma: A Review of the North American Clinical Trial Perspective. *Front. Oncol.* **2018**, *8*, 169. [CrossRef] [PubMed]

37. Gaspar, N.; Marshall, L.; Perryman, L.; Bax, D.A.; Little, S.E.; Viana-Pereira, M.; Sharp, S.Y.; Vassal, G.; Pearson, A.D.; Reis, R.M.; et al. MGMT-independent temozolomide resistance in pediatric glioblastoma cells associated with a PI3-kinase-mediated HOX/stem cell gene signature. *Cancer Res.* **2010**, *70*, 9243–9252. [CrossRef] [PubMed]

38. Kurscheid, S.; Bady, P.; Sciuscio, D.; Samarzija, I.; Shay, T.; Vassallo, I.; Criekinge, W.V.; Daniel, R.T.; van den Bent, M.J.; Marosi, C.; et al. Chromosome 7 gain and DNA hypermethylation at the HOXA10 locus are associated with expression of a stem cell related HOX-signature in glioblastoma. *Genome Biol.* **2015**, *16*, 16. [CrossRef] [PubMed]

39. Costa, B.M.; Smith, J.S.; Chen, Y.; Chen, J.; Phillips, H.S.; Aldape, K.D.; Zardo, G.; Nigro, J.; James, C.D.; Fridlyand, J.; et al. Reversing HOXA9 oncogene activation by PI3K inhibition: Epigenetic mechanism and prognostic significance in human glioblastoma. *Cancer Res.* **2010**, *70*, 453–462. [CrossRef] [PubMed]

40. Arrowsmith, C.H.; Audia, J.E.; Austin, C.; Baell, J.; Bennett, J.; Blagg, J.; Bountra, C.; Brennan, P.E.; Brown, P.J.; Bunnage, M.E.; et al. The promise and peril of chemical probes. *Nat. Chem. Biol.* **2015**, *11*, 536–541. [CrossRef] [PubMed]

41. Gharbi, S.I.; Zvelebil, M.J.; Shuttleworth, S.J.; Hancox, T.; Saghir, N.; Timms, J.F.; Waterfield, M.D. Exploring the specificity of the PI3K family inhibitor LY294002. *Biochem. J.* **2007**, *404*, 15–21. [CrossRef] [PubMed]

42. Dittmann, A.; Werner, T.; Chung, C.W.; Savitski, M.M.; Falth Savitski, M.; Grandi, P.; Hopf, C.; Lindon, M.; Neubauer, G.; Prinjha, R.K.; et al. The commonly used PI3-kinase probe LY294002 is an inhibitor of BET bromodomains. *ACS Chem. Biol.* **2014**, *9*, 495–502. [CrossRef] [PubMed]

43. Wu, Y.L.; Maachani, U.B.; Schweitzer, M.; Singh, R.; Wang, M.; Chang, R.; Souweidane, M.M. Dual Inhibition of PI3K/AKT and MEK/ERK Pathways Induces Synergistic Antitumor Effects in Diffuse Intrinsic Pontine Glioma Cells. *Transl. Oncol.* **2017**, *10*, 221–228. [CrossRef] [PubMed]

44. Miyahara, H.; Yadavilli, S.; Natsumeda, M.; Rubens, J.A.; Rodgers, L.; Kambhampati, M.; Taylor, I.C.; Kaur, H.; Asnaghi, L.; Eberhart, C.G.; et al. The dual mTOR kinase inhibitor TAK228 inhibits tumorigenicity and enhances radiosensitization in diffuse intrinsic pontine glioma. *Cancer Lett.* **2017**, *400*, 110–116. [CrossRef] [PubMed]

45. Venkatesh, H.S.; Tam, L.T.; Woo, P.J.; Lennon, J.; Nagaraja, S.; Gillespie, S.M.; Ni, J.; Duveau, D.Y.; Morris, P.J.; Zhao, J.J.; et al. Targeting neuronal activity-regulated neuroligin-3 dependency in high-grade glioma. *Nature* **2017**, *549*, 533–537. [CrossRef] [PubMed]

46. Becher, O.J.; Gilheeney, S.W.; Khakoo, Y.; Lyden, D.C.; Haque, S.; De Braganca, K.C.; Kolesar, J.M.; Huse, J.T.; Modak, S.; Wexler, L.H.; et al. A phase I study of perifosine with temsirolimus for recurrent pediatric solid tumors. *Pediatr. Blood Cancer* **2017**, *64*. [CrossRef] [PubMed]

47. Nikbakht, H.; Panditharatna, E.; Mikael, L.G.; Li, R.; Gayden, T.; Osmond, M.; Ho, C.Y.; Kambhampati, M.; Hwang, E.I.; Faury, D.; et al. Spatial and temporal homogeneity of driver mutations in diffuse intrinsic pontine glioma. *Nat. Commun.* **2016**, *7*, 11185. [CrossRef] [PubMed]
48. Bjerke, L.; Mackay, A.; Nandhabalan, M.; Burford, A.; Jury, A.; Popov, S.; Bax, D.A.; Carvalho, D.; Taylor, K.R.; Vinci, M.; et al. Histone H3.3. mutations drive pediatric glioblastoma through upregulation of MYCN. *Cancer Discov.* **2013**, *3*, 512–519. [CrossRef] [PubMed]
49. Kozono, D.; Li, J.; Nitta, M.; Sampetrean, O.; Gonda, D.; Kushwaha, D.S.; Merzon, D.; Ramakrishnan, V.; Zhu, S.; Zhu, K.; et al. Dynamic epigenetic regulation of glioblastoma tumorigenicity through LSD1 modulation of MYC expression. *Proc. Natl. Acad. Sci. USA* **2015**, *112*, E4055–E4064. [CrossRef] [PubMed]
50. Funato, K.; Major, T.; Lewis, P.W.; Allis, C.D.; Tabar, V. Use of human embryonic stem cells to model pediatric gliomas with H3.3K27M histone mutation. *Science* **2014**, *346*, 1529–1533. [CrossRef] [PubMed]
51. Grembecka, J.; He, S.; Shi, A.; Purohit, T.; Muntean, A.G.; Sorenson, R.J.; Showalter, H.D.; Murai, M.J.; Belcher, A.M.; Hartley, T.; et al. Menin-MLL inhibitors reverse oncogenic activity of MLL fusion proteins in leukemia. *Nat. Chem. Biol.* **2012**, *8*, 277–284. [CrossRef] [PubMed]
52. Hashizume, R.; Andor, N.; Ihara, Y.; Lerner, R.; Gan, H.; Chen, X.; Fang, D.; Huang, X.; Tom, M.W.; Ngo, V.; et al. Pharmacologic inhibition of histone demethylation as a therapy for pediatric brainstem glioma. *Nat. Med.* **2014**, *20*, 1394–1396. [CrossRef] [PubMed]
53. Piunti, A.; Hashizume, R.; Morgan, M.A.; Bartom, E.T.; Horbinski, C.M.; Marshall, S.A.; Rendleman, E.J.; Ma, Q.; Takahashi, Y.H.; Woodfin, A.R.; et al. Therapeutic targeting of polycomb and BET bromodomain proteins in diffuse intrinsic pontine gliomas. *Nat. Med.* **2017**, *23*, 493–500. [CrossRef] [PubMed]
54. Nagaraja, S.; Vitanza, N.A.; Woo, P.J.; Taylor, K.R.; Liu, F.; Zhang, L.; Li, M.; Meng, W.; Ponnuswami, A.; Sun, W.; et al. Transcriptional Dependencies in Diffuse Intrinsic Pontine Glioma. *Cancer Cell* **2017**, *31*, 635–652. [CrossRef] [PubMed]
55. Zhang, Y.; Dong, W.; Zhu, J.; Wang, L.; Wu, X.; Shan, H. Combination of EZH2 inhibitor and BET inhibitor for treatment of diffuse intrinsic pontine glioma. *Cell Biosci.* **2017**, *7*, 56. [CrossRef] [PubMed]
56. Mohammad, F.; Weissmann, S.; Leblanc, B.; Pandey, D.P.; Hojfeldt, J.W.; Comet, I.; Zheng, C.; Johansen, J.V.; Rapin, N.; Porse, B.T.; et al. EZH2 is a potential therapeutic target for H3K27M-mutant pediatric gliomas. *Nat. Med.* **2017**, *23*, 483–492. [CrossRef] [PubMed]
57. Cordero, F.J.; Huang, Z.; Grenier, C.; He, X.; Hu, G.; McLendon, R.E.; Murphy, S.K.; Hashizume, R.; Becher, O.J. Histone H3.3K27M Represses p16 to Accelerate Gliomagenesis in a Murine Model of DIPG. *Mol. Cancer Res.* **2017**, *15*, 1243–1254. [CrossRef] [PubMed]
58. Halvorson, K.G.; Barton, K.L.; Schroeder, K.; Misuraca, K.L.; Hoeman, C.; Chung, A.; Crabtree, D.M.; Cordero, F.J.; Singh, R.; Spasojevic, I.; et al. A high-throughput in vitro drug screen in a genetically engineered mouse model of diffuse intrinsic pontine glioma identifies BMS-754807 as a promising therapeutic agent. *PLoS ONE* **2015**, *10*, e0118926. [CrossRef] [PubMed]
59. Kumar, S.S.; Sengupta, S.; Lee, K.; Hura, N.; Fuller, C.; DeWire, M.; Stevenson, C.B.; Fouladi, M.; Drissi, R. BMI-1 is a potential therapeutic target in diffuse intrinsic pontine glioma. *Oncotarget* **2017**, *8*, 62962–62975. [CrossRef] [PubMed]
60. Filbin, M.G.; Tirosh, I.; Hovestadt, V.; Shaw, M.L.; Escalante, L.E.; Mathewson, N.D.; Neftel, C.; Frank, N.; Pelton, K.; Hebert, C.M.; et al. Developmental and oncogenic programs in H3K27M gliomas dissected by single-cell RNA-seq. *Science* **2018**, *360*, 331–335. [CrossRef] [PubMed]
61. Zuckermann, M.; Hovestadt, V.; Knobbe-Thomsen, C.B.; Zapatka, M.; Northcott, P.A.; Schramm, K.; Belic, J.; Jones, D.T.; Tschida, B.; Moriarity, B.; et al. Somatic CRISPR/Cas9-mediated tumour suppressor disruption enables versatile brain tumour modelling. *Nat. Commun.* **2015**, *6*, 7391. [CrossRef] [PubMed]
62. Herz, H.M.; Morgan, M.; Gao, X.; Jackson, J.; Rickels, R.; Swanson, S.K.; Florens, L.; Washburn, M.P.; Eissenberg, J.C.; Shilatifard, A. Histone H3 lysine-to-methionine mutants as a paradigm to study chromatin signaling. *Science* **2014**, *345*, 1065–1070. [CrossRef] [PubMed]
63. Zingg, D.; Debbache, J.; Peña-Hernández, R.; Antunes, A.T.; Schaefer, S.M.; Cheng, P.F.; Zimmerli, D.; Haeusel, J.; Calçada, R.R.; Tuncer, E.; et al. EZH2-Mediated Primary Cilium Deconstruction Drives Metastatic Melanoma Formation. *Cancer Cell* **2018**, *34*. [CrossRef] [PubMed]
64. Nakamura, Y.; Hattori, N.; Iida, N.; Yamashita, S.; Mori, A.; Kimura, K.; Yoshino, T.; Ushijima, T. Targeting of super-enhancers and mutant BRAF can suppress growth of BRAF-mutant colon cancer cells via repression of MAPK signaling pathway. *Cancer Lett.* **2017**, *402*, 100–109. [CrossRef] [PubMed]

65. Zakharia, Y.; Monga, V.; Swami, U.; Bossler, A.D.; Freesmeier, M.; Frees, M.; Khan, M.; Frydenlund, N.; Srikantha, R.; Vanneste, M.; et al. Targeting epigenetics for treatment of BRAF mutated metastatic melanoma with decitabine in combination with vemurafenib: A phase lb study. *Oncotarget* **2017**, *8*, 89182–89193. [CrossRef] [PubMed]

66. Grasso, C.S.; Tang, Y.; Truffaux, N.; Berlow, N.E.; Liu, L.; Debily, M.A.; Quist, M.J.; Davis, L.E.; Huang, E.C.; Woo, P.J.; et al. Functionally defined therapeutic targets in diffuse intrinsic pontine glioma. *Nat. Med.* **2015**, *21*, 555–559. [CrossRef] [PubMed]

67. Brown, Z.Z.; Muller, M.M.; Jain, S.U.; Allis, C.D.; Lewis, P.W.; Muir, T.W. Strategy for "detoxification" of a cancer-derived histone mutant based on mapping its interaction with the methyltransferase PRC2. *J. Am. Chem. Soc.* **2014**, *136*, 13498–13501. [CrossRef] [PubMed]

68. Singh, M.M.; Manton, C.A.; Bhat, K.P.; Tsai, W.W.; Aldape, K.; Barton, M.C.; Chandra, J. Inhibition of LSD1 sensitizes glioblastoma cells to histone deacetylase inhibitors. *Neuro Oncol.* **2011**, *13*, 894–903. [CrossRef] [PubMed]

69. Singh, M.M.; Johnson, B.; Venkatarayan, A.; Flores, E.R.; Zhang, J.; Su, X.; Barton, M.; Lang, F.; Chandra, J. Preclinical activity of combined HDAC and KDM1A inhibition in glioblastoma. *Neuro Oncol.* **2015**, *17*, 1463–1473. [CrossRef] [PubMed]

70. Drappatz, J.; Lee, E.Q.; Hammond, S.; Grimm, S.A.; Norden, A.D.; Beroukhim, R.; Gerard, M.; Schiff, D.; Chi, A.S.; Batchelor, T.T.; et al. Phase I study of panobinostat in combination with bevacizumab for recurrent high-grade glioma. *J. Neurooncol.* **2012**, *107*, 133–138. [CrossRef] [PubMed]

71. Lee, E.Q.; Reardon, D.A.; Schiff, D.; Drappatz, J.; Muzikansky, A.; Grimm, S.A.; Norden, A.D.; Nayak, L.; Beroukhim, R.; Rinne, M.L.; et al. Phase II study of panobinostat in combination with bevacizumab for recurrent glioblastoma and anaplastic glioma. *Neuro Oncol.* **2015**, *17*, 862–867. [CrossRef] [PubMed]

72. Cooney, T.; Onar-Thomas, A.; Huang, J.; Lulla, R.; Fangusaro, J.; Kramer, K.; Baxter, P.; Fouladi, M.; Dunkel, I.J.; Warren, K.E.; et al. Dipg-22. A phase 1 trial of the histone deacetylase inhibitor panobinostat in pediatric patients with recurrent or refractory diffuse intrinsic pontine glioma: A pediatric brain tumor consortium (pbtc) study. *Neuro Oncol.* **2018**, *20*. [CrossRef]

73. Masoudi, A.; Elopre, M.; Amini, E.; Nagel, M.E.; Ater, J.L.; Gopalakrishnan, V.; Wolff, J.E. Influence of valproic acid on outcome of high-grade gliomas in children. *Anticancer Res.* **2008**, *28*, 2437–2442. [PubMed]

74. Subashi, E.; Cordero, F.J.; Halvorson, K.G.; Qi, Y.; Nouls, J.C.; Becher, O.J.; Johnson, G.A. Tumor location, but not H3.3K27M, significantly influences the blood-brain-barrier permeability in a genetic mouse model of pediatric high-grade glioma. *J. Neurooncol.* **2016**, *126*, 243–251. [CrossRef] [PubMed]

75. Wei, K.C.; Chu, P.C.; Wang, H.Y.; Huang, C.Y.; Chen, P.Y.; Tsai, H.C.; Lu, Y.J.; Lee, P.Y.; Tseng, I.C.; Feng, L.Y.; et al. Focused ultrasound-induced blood-brain barrier opening to enhance temozolomide delivery for glioblastoma treatment: A preclinical study. *PLoS ONE* **2013**, *8*, e58995. [CrossRef] [PubMed]

76. Fan, C.H.; Ting, C.Y.; Liu, H.L.; Huang, C.Y.; Hsieh, H.Y.; Yen, T.C.; Wei, K.C.; Yeh, C.K. Antiangiogenic-targeting drug-loaded microbubbles combined with focused ultrasound for glioma treatment. *Biomaterials* **2013**, *34*, 2142–2155. [CrossRef] [PubMed]

77. Peiris, P.M.; Abramowski, A.; McGinnity, J.; Doolittle, E.; Toy, R.; Gopalakrishnan, R.; Shah, S.; Bauer, L.; Ghaghada, K.B.; Hoimes, C.; et al. Treatment of Invasive Brain Tumors Using a Chain-like Nanoparticle. *Cancer Res.* **2015**, *75*, 1356–1365. [CrossRef] [PubMed]

78. Absinta, M.; Ha, S.K.; Nair, G.; Sati, P.; Luciano, N.J.; Palisoc, M.; Louveau, A.; Zaghloul, K.A.; Pittaluga, S.; Kipnis, J.; et al. Human and nonhuman primate meninges harbor lymphatic vessels that can be visualized noninvasively by MRI. *Elife* **2017**, *6*. [CrossRef] [PubMed]

79. Chheda, Z.S.; Kohanbash, G.; Okada, K.; Jahan, N.; Sidney, J.; Pecoraro, M.; Yang, X.; Carrera, D.A.; Downey, K.M.; Shrivastav, S.; et al. Novel and shared neoantigen derived from histone 3 variant H3.3K27M mutation for glioma T cell therapy. *J. Exp. Med.* **2018**, *215*, 141–157. [CrossRef] [PubMed]

80. Mount, C.W.; Majzner, R.G.; Sundaresh, S.; Arnold, E.P.; Kadapakkam, M.; Haile, S.; Labanieh, L.; Hulleman, E.; Woo, P.J.; Rietberg, S.P.; et al. Potent antitumor efficacy of anti-GD2 CAR T cells in H3-K27M(+) diffuse midline gliomas. *Nat. Med.* **2018**, *24*, 572–579. [CrossRef] [PubMed]

81. Richman, S.A.; Nunez-Cruz, S.; Moghimi, B.; Li, L.Z.; Gershenson, Z.T.; Mourelatos, Z.; Barrett, D.M.; Grupp, S.A.; Milone, M.C. High-Affinity GD2-Specific CAR T Cells Induce Fatal Encephalitis in a Preclinical Neuroblastoma Model. *Cancer Immunol. Res.* **2018**, *6*, 36–46. [CrossRef] [PubMed]

82. Bailey, C.; Romero, M.; Han, R.; Larson, J.; Becher, O.; Lee, D.; Monje, M.; Gopalakrishnan, V.; Zaky, W.; Chandra, J. Immu-19. LSD1 modulates nk cell immunotherapy through an onco-immunogenic gene signature in dipg. *Neuro Oncol.* **2018**, *20*. [CrossRef]

83. Reid, M.A.; Dai, Z.; Locasale, J.W. The impact of cellular metabolism on chromatin dynamics and epigenetics. *Nat. Cell Biol.* **2017**, *19*, 1298–1306. [CrossRef] [PubMed]

84. Baron, O.; Förthmann, B.; Lee, Y.W.; Terranova, C.; Ratzka, A.; Stachowiak, E.K.; Grothe, C.; Claus, P.; Stachowiak, M.K. Cooperation of nuclear fibroblast growth factor receptor 1 and Nurr1 offers new interactive mechanism in postmitotic development of mesencephalic dopaminergic neurons. *J. Biol. Chem.* **2012**, *287*, 19827–19840. [CrossRef] [PubMed]

85. Choubey, L.; Collette, J.C.; Smith, K.M. Quantitative assessment of fibroblast growth factor receptor 1 expression in neurons and glia. *Peerj* **2017**, *5*, e3173. [CrossRef] [PubMed]

86. Funa, K.; Sasahara, M. The roles of PDGF in development and during neurogenesis in the normal and diseased nervous system. *J. Neuroimmune Pharmacol.* **2014**, *9*, 168–181. [CrossRef] [PubMed]

87. Miyata, T.; Toho, T.; Nonoguchi, N.; Furuse, M.; Kuwabara, H.; Yoritsune, E.; Kawabata, S.; Kuroiwa, T.; Miyatake, S. The roles of platelet-derived growth factors and their receptors in brain radiation necrosis. *Radiat. Oncol.* **2014**, *9*, 51. [CrossRef] [PubMed]

88. Pacifici, M.; Shore, E.M. Common mutations in ALK2/ACVR1, a multi-faceted receptor, have roles in distinct pediatric musculoskeletal and neural orphan disorders. *Cytokine Growth Factor Rev.* **2016**, *27*, 93–104. [CrossRef] [PubMed]

© 2018 by the authors. Licensee MDPI, Basel, Switzerland. This article is an open access article distributed under the terms and conditions of the Creative Commons Attribution (CC BY) license (http://creativecommons.org/licenses/by/4.0/).

bioengineering

MDPI

Review

Palliative Care for Children with Central Nervous System Malignancies

Peter H. Baenziger [1] and Karen Moody [2,*]

[1]   Peyton Manning Children's Hospital, Ascension St. Vincent, 2001 West 86th Street, Indianapolis, IN 46260, USA; peter.baenziger@ascension.org
[2]   MD Anderson Cancer Center, University of Texas, 1515 Holcomb Blvd., Unit 87, Houston, TX 77030, USA
*   Correspondence: Kmoody@mdanderson.org

Received: 15 August 2018; Accepted: 3 October 2018; Published: 13 October 2018

Abstract: Children with central nervous system (CNS) malignancies often suffer from high symptom burden and risk of death. Pediatric palliative care is a medical specialty, provided by an interdisciplinary team, which focuses on enhancing quality of life and minimizing suffering for children with life-threatening or life-limiting disease, and their families. Primary palliative care skills, which include basic symptom management, facilitation of goals-of-care discussions, and transition to hospice, can and should be developed by all providers of neuro-oncology care. This chapter will review the fundamentals of providing primary pediatric palliative care.

Keywords: palliative care; child; brain; neoplasm; neuropathic pain; pain; symptoms; hospice

---

1. Introduction

Neurological tumors comprise the second most common malignancy of children and adolescents, with an incidence of approximately 5000 children affected per year [1]. In addition, neurological tumors represent the most common cause of cancer-related death in children and adolescents [2]. Neurological tumors generally fall into high or low grade strata and generally higher grade tumors have poorer survival, however, patients with either high grade or low grade tumors suffer the adverse effects of chemotherapy, radiation, surgery, and direct disease sequelae, which may be ameliorated with palliative care.

Palliative care for children is critical to providing optimal care for some of the most vulnerable patients [3,4]. Providers caring for children with life-threatening illness should have a fundamental understanding of how to assist patients and families in establishing goals-of-care and essential pain and symptom management skills, the mainstays of the field of hospice and palliative medicine [5–7]. The American Academy of Pediatrics outlines principles to guide palliative care practice, including: (1) Providers have an obligation to ensure interventions are only used when potential benefits outweigh risks; (2) the goal of palliative care is to enhance quality of life despite the disease trajectory; (3) palliative care focuses on symptoms and conditions; and (4) palliative care teams work towards healthy bereavement for the family of the patient. Palliative care is generally provided by multidisciplinary teams to address the wide variety of burdens faced by patients suffering from serious illness, including neurologic tumors [8–10]. There is consensus that the ideal team includes a physician, nurse, social worker, spiritual advisor, and a child life therapist [5].

2. Facilitating Discussion and Decisions

A critical aspect of primary palliative care is to assist with establishing goals-of-care with patients and families. Establishing goals first requires some determination of the patient's and family's values (herein, simply "the family") and priorities; this is achieved across a series of conversations. Examples of common values and priorities include: Staying out of the hospital, controlling pain, spending time with loved ones, and achieving milestones, such as graduation [11,12]. These values are discussed in the context of the patient's illness, treatment options, and prognosis to establish priorities and goals of care and to facilitate medical decision-making. The medical decisions faced by parents for children with brain tumors are often numerous, frequent, high-stakes decisions and should align with these established goals of care. A key to facilitating these difficult conversations is to communicate in a way that "meets people where they are"—essentially, by not giving too much or too little information (which can be assessed through direct inquiry) and by enabling the patient's and family's values to drive the conversation. Goal setting is best accomplished when there is a consideration of the broader view of a patient and family, beyond the disease context, to include their personal lives, their community, and their social, emotional, and spiritual wellness [13].

Communicating well is important to patients, decreases uncertainty, and may improve hope and reduce decision regret [14]. Most parents would prefer to receive information about palliative care treatment options for their children than for this information to be withheld [14]. Young adults with cancer who have experienced an honest prognostic discussion with their providers demonstrate increased trust in providers, increased peace and hope, and decreased distress [15]. See Table 1 for communication tools [16].

Table 1. Palliative care communication tools.

Delivering Bad News	Verbal Responses to Emotion	Non-Verbal Response to Emotion
"SPIKES": [17] -Setting: Prepare the setting for the conversation and minimize distractions. -Perception: Assess the caregiver's perception of the clinical information. -Invitation: Ask permission to deliver new information. -Knowledge: Provide the main message up front, simply. -Emotion: Respond empathically to emotion. -Strategy/Summary: Summarize the encounter and what will happen next.	"NURSES": [18,19] -Naming the emotion statement: "I hear frustration in your voice." -Understanding statement: "I understand this is upsetting news." -Respect statement: "I can see how dedicated you have been to your son's care over these three months." -Support statement: "We are here to help you and your family." -Explore statement: "Tell me about what you were hoping to hear today." -Silence: Providing silence in the room can passively, yet explicitly recognize emotions.	"SOLAR": [20] -Squarely face the patient. -Open body posture. -Lean towards the patient. -Eye contact. -Relaxed body posture.

Palliative care principles obligate providers and families to provide a developmentally appropriate explanation of disease and the burdens expected to the patient within the context of their values, culture, and preferences [5]. Earlier discussion of prognosis may be beneficial to allow for processing of information and allowing for earlier integration of prognosis into decision making [10]. Intentional, compassionate discussions of death should include consideration of the child's developmental maturity, understanding of death, prior experiences with death, and the cultural and religious milieu in which the child lives, including family preference about if, when, and how information is disclosed to the patient [5]. Families should be encouraged to discuss the child's fears, meeting them at a developmentally appropriate level [20,21]. It is important to note that studies demonstrate children are often aware of their grave prognosis before parents or care teams engage in a formal conversation and that children may avoid the topic to protect their parents [22].

A common decision put to families is whether and when to enroll in hospice. Hospice is a comprehensive home care program, with 24-hour, on-call nurse visits, home medication delivery, continuous 1:1 nursing if needed, and bereavement support. Eligibility for hospice only requires a reasonable possibility of death within six months; a do not resuscitation order is not required. Hospice utilizes a multidisciplinary approach focusing on comfort and quality of life. Advanced care planning is important for end-of-life care, including the patient's/family's preference for location of death.

Unique to pediatrics is the concept of "concurrent care", which allows children to "concurrently" receive both curative therapies and hospice [23].

Another decision many families face is Phase I trial enrollment, which may provide families with a sense of hope, legacy, and dignity, despite a minimal likelihood of objective disease response. The burdens of the trial should be balanced with the family's sense of benefit. Theoretically, enrollment on a clinical trial near end of life can complicate comfort-directed care plans because of study-related requirements, however, available data suggest that patients that enroll on Phase I trials have similar timing and frequencies of "do not resuscitate" (DNR) discussions, hospice enrollment, and established DNR orders to those that do not enroll on such trials [24].

Since medical decisions in the palliative setting are value-based, conflicts may arise between family decision makers and medical teams regarding the goals of care and specifics of treatment. Conflict can be addressed at both a provider level and an institutional level. Providers of palliative care can mitigate disagreement by "shifting" to a more curious stance, in which "one focus[es] on learning more about the perspectives of both parties, exploring what may be a complex web of actions that contributed to the conflict." [25]. A simple example of one of the ways one can make such a shift is repeating back what one hears a family member express as their hope; "so I hear you saying you hope there is a surgery that will stop your son's cancer". Identifying a common ground among identified hopes of family members is a helpful first step in negotiating a resolution. Many other practical communication tools have been summarized and are worthy of further reading and practice [26]. At an institution level, it is important to have a process of conflict resolution [27]. Such policies usually include purposeful communication efforts, opportunities for families to seek second opinions, and consultation with an ethics committee.

Symptom Burden

The symptoms experienced by children with brain tumors are varied and can be quite burdensome [8]. The location of the tumor may predict likely burdens: Supratentorial tumors are associated with seizures, coma, and nausea and vomiting; infratentorial tumors are associated with ataxia; brain stem tumors are associated with speech disturbances, cranial nerve paralysis (swallowing dysfunction), and tetraparesis [28,29]. In addition to the most common symptoms of other childhood cancers (fatigue, pain, dyspnea), children with brain tumors can also suffer dysphagia and dysarthria, hearing and vision loss, paralysis, seizures, agitation, headaches, and cognitive and behavioral changes [30–32].

3. Oromotor Dysfunction and Secretions

Expert experience provides the bulk of treatment recommendations for oromotor dysfunction, which include speech therapy evaluation, thickening of feeds, and treatment with steroids to acutely reduce edema.

Secretions often become difficult for patients to manage as they lose their oral motor proficiency and their alert mental status. The goal of treatment is to decrease respiratory distress, aesthetic distress of secretions, and to attain maximum patient comfort. Medicines used are muscarinic anticholinergics, such as glycopyrrolate, scopolamine, or atropine. Glycopyrrolate does not cross the blood-brain barrier and thus spares patients from central nervous system side effects. Scopolamine patches pose a convenient transdermal delivery system and are approved for patients greater than 45 kg. Atropine ophthalmic 1% solution can be given sublingually; 1 drop every 4 h providing 0.5 mg atropine per dose [33]. Other treatment strategies are regular oral care and suction, lateral positioning of the patient, and music to dampen the noise. Families can be reassured that the rattling noise is not painful (an analogy to snoring works well) and deep suction will generally cause more harm than benefit.

Nutrition and Hydration

As oromotor dysfunction progresses to dysphagia, nutrition and hydration often become concerns that need particular attention due to the fundamental nature of care through feeding. Nutrition and hydration should be framed by prognosis, risks (such as aspiration pneumonia), benefits (child's enjoyment of favorite foods, tastes, and experiences; social interactions surrounding meals), and focused on the experience of the patient. For those patients who can no longer feed by mouth yet express hunger or other symptoms due to a lack of intake, the benefits to intervening may outweigh the risks. Generally, if a child is likely to live longer than 90 days, it is reasonable to use more invasive solutions (surgically placed feeding tube, total parenteral nutrition) if such procedures are aligned with family and patient values. If the prognosis is shorter, a nasogastric or nasojejunal feeding tube is more appropriate, with decreased risks, and foregoing feeds and fluids altogether is also an acceptable option for some. The burdens of feeds should not be forgotten; patients may feel discomfort from the feed itself, the tubes, or may have adverse responses to their change in appearance. When the primary goal is patient comfort, even when oral feeding poses an aspiration risk, it is reasonable to allow small amounts of intake for enjoyment. Conversely, when the primary goal is patient comfort, it is ethically sound to forego nutrition and hydration that are burdensome to the patient. Finally, offering hydration without nutrition or in the setting of impending death, often leads to increased symptom burden (dyspnea, swelling, pain) compared to mild dehydration [34].

4. Communication Difficulties

Children with neurologic tumors may suffer frustrating decreases in their ability to communicate due to dysarthria, aphasia, or hearing deficits [35,36]. Assisted communication devices require significant time to learn, thus they are less useful than simple tools, like dry erase boards or symbol books [37–39]. Speech, occupational, and physical therapies can have a significant impact by slowing the decline of key functional quality of life skills, such as speech and swallowing [8,40].

5. Headache

Headache occurs in around 36% of patients with brain tumors [41]. The primary treatment modality is pharmacologic, including acetaminophen, opioids, and dexamethasone. Medications that address neuropathic pain can be beneficial, but often take a week or more to be effective. Low dose methadone (0.04 mg/kg/dose BID) is an opioid that also has N-Methyl-D-aspartic acid (NMDA) receptor antagonist properties and can be particularly efficacious in cancer related headache, but may necessitate consultation with a palliative care or pain specialist comfortable with dosing and monitoring [42]. Methadone can be compounded into a 10 mg per mL solution for easy sublingual administration near end of life. Morphine can be similarly compounded into highly concentrated sublingual solutions. When using steroids for headache, the lowest possible effective dose should be used, and weaning should be considered after control of symptoms. NSAIDs (non-steroidal anti-inflammatory drugs) are typically avoided due to risk of bleeding at the tumor site and due to the risk of gastric ulcers with concurrent use of steroids.

6. Seizures

Seizures occur in 30–50% of patients with brain tumors. Seizures cause distress to the patient and family and may also cause worsening cognitive and behavioral functioning [43,44]. The treatment goals are to control seizures and to optimize periods of alertness. There is no evidence to support routine prophylactic antiepileptic use in the absence of a documented seizure history and these agents may cause sedation or agitation [45]. For acute seizure treatment, the benzodiazepines are most helpful. Diazepam and lorazepam can be given rectally, and lorazepam and midazolam are both absorbed intranasally [45].

For chronic seizure suppression, levetiracetam, as a broad acting staple, has few adverse effects, is well tolerated, and is often effective [46]. Many patients experience a brief, initial period of aggression on levetiracetam; if it does not resolve within several weeks, vitamin B6 supplementation has been shown in a small study to reduce the behavior change [47]. Pediatric neurology consultation is prudent for refractory seizures, status epilepticus, or patient intolerance due to adverse effects.

Because neurologic tumors are associated with dysphagia, concentrated sublingual compounds and rectally administered antiepileptic medications often become necessary. Pharmacists are an important resource in this regard. Diazepam is the primary rectal choice as it comes in a suppository and acts more quickly than other antiepileptics given by this route.

7. Nausea and Vomiting

Nausea and vomiting in the setting of pediatric neural tumors may be due to treatment or direct effects of the tumor. The vomit center receives input from two central nervous system controlling areas (chemoreceptor trigger zone; nucleus tractus solitarius) [48] and one peripheral input source (vagus nerve) [49], which use a multitude of neurotransmitters (histamine, acetylcholine, dopamine, serotonin, neurokinin) [50]. Treatment goals are to reduce nausea and vomiting to the point that patients can interact with family, can take by mouth foods they enjoy, and maintain nutrition and hydration.

Consider whether nausea is due to direct pressure from tumor, chemotherapy, toxins (uremia, hepatic failure, hypercalcemia), vestibular pathology, or gastrointestinal pathology (obstruction, ulcers, mucositis, constipation) [48]. When the cause is increased intracranial pressure, steroids are a first-line treatment. When chemotherapy or radiation-induced nausea and vomiting is expected, prophylaxis is recommended, and several published guidelines for children exist [51,52]. For the best control of nausea and vomiting, one should utilize both a scheduled medication as well as an as-needed medication; choosing medications from different classes is also ideal. See Table 2 for a list of antiemetics and dosages. Non-medical therapies can include hypnosis [53,54] and acupuncture [55]. Additionally, practical remedies, such as treating constipation, removing noxious smells, opioid rotation, emotional support [56], and modifying meals to be smaller and more frequent, can be helpful [55].

Table 2. Antiemetics.

Class	Drug	Dose	Forms	Notes
NK-1 Antagonist	Aprepitant (Emend)	Day 1: 3 mg/kg PO (max 125 mg) Day 2, 3: 2 mg/kg PO (max 80 mg)	Capsule, suspension	Approved for chemotherapy induced nausea/vomiting (CINV). Assess for CYP3A4 & 2C9 drug interactions. Minimal data exists on the use of fosaprepitant in children <12 years
Steroid	Dexamethasone (Decadron)	10 mg/m² IV/PO daily (reduce to 5 mg/m² if using with aprepitant)	IV, tablet, solution	This is the CINV dose; alternate dosing is used for brain edema
5HT3 Antagonist	Ondansetron (Zofran)	0.15 mg/kg/dose IV/PO q8 hours (max 8 mg/dose)	IV, tablet, oral disintegrating tablet, solution	5HT3 antagonists have equivalent efficacy at comparable doses
	Granisetron (Kytril)	0.04 mg/kg IV daily or PO q12 hours (max 1 mg/dose) age >12 years: 1–2 mg PO/IV q12 hours	IV, tablet, solution (custom compounded), patch (available as outpatient prescription for adolescents)	
	Palonosetron (Aloxi)	0.02 mg/kg IV once prior to chemo. If necessary, may re-dose 72 hours later	IV	

Table 2. *Cont.*

Class	Drug	Dose	Forms	Notes
Phenothiazine	Promethazine (Phenergan)	0.25 mg/kg PO/IV q6 hours (max 25 mg/dose)	IV, tablet, syrup, suppository, topical gel	Contraindicated in children <2 years old. Anticholinergic.
	Prochlorperazine (Compazine)	0.1 mg/kg/dose IV/PO q6 hours (max 10 mg/dose)	IV, tablet, suppository	Contraindicated in children <2 years old or <9 kg; anticholinergic and anti-dopaminergic; risk of extrapyramidal symptoms
Prokinetic	Metoclopramide (Reglan)	0.1–0.5 mg/kg IV/PO q6 hours (max 10 mg) Adolescents: 5–10 mg IV/PO q6 hours	IV, tablet, suspension	Risk of tardive dyskinesia, especially with prolonged use; may use with oral diphenhydramine. Anti-dopaminergic
Benzodiazepine	Lorazepam (Ativan)	0.04 mg/kg IV/PO q8 hours (max 2 mg/dose)	IV, tablet, suspension	Risk of sedation, respiratory depression, coma, and death when used with opioids
Atypical Antipsychotic	Olanzapine (Zyprexa)	0.14 mg/kg/dose PO qHS (max 5–10 mg/dose)	tablet, orally disintegrating tablet	Antidopaminergic, anticholinergic, and 5HT2 antagonist.
Cannabinoid	Dronabinol (Marinol)	5 mg/m^2 PO BID-QID (max 10 mg/dose)	Capsule	Contraindicated with sesame oil hypersensitivity
Antihistamine	Diphenhydramine (Benadryl)	0.5 mg/kg PO q6 hours	Oral, elixir	Avoid IV use due to dependency and sedation risk. Also, may use to manage EPS side effects.
Anticholinergic	Scopolamine (Transderm Scop)	1.5 mg patch changed q72 hours	Patch	For use in patients >45 kg
Butyrophenone	Haloperidol (Haldol)	3–12 years old start 0.05 mg/kg/day divided BID-TID >12 start 0.5 mg per dose BID-TID, up to 4 mg/dose q 6 hours	PO tabs/IV/SC	Anti-dopaminergic. Risk of severe extrapyramidal symptoms, prolonged QT and granulocytopenia

8. Pain

Pain is reported by many children and adolescents with brain tumors, and may be due to the cancer itself, the treatments received, and the procedures performed [33]. Most patients with neurological tumors benefit from pain medicines at some point in their disease process (84%) [57]. Neuropathic pain caused by the tumor can be particularly challenging to treat and may require polypharmacy for best results.

Other manifestations of "total pain" [58] are addressed hereafter, including psychological, social, emotional, and spiritual elements [59]. These must be treated in parallel to physical pain as they are codependent.

Treatment of pain in the patient with neuro-oncological disease begins with the foundations of pain in other clinical care arenas: Assessment, identifying type of pain, intervention, and reassessment. Validated scales should be used whenever possible to track progress. The FLACC scale [60] is useful for caregiver observation of infants and other non-verbal patients. For children 3 years and older, the Wong Baker FACES scale gives an age appropriate visual scale and, for ages 9 years and older [61], a simple 0–10 numerical rating scale is appropriate.

The two fundamental types of pain are nociceptive and neuropathic. Nociceptive pain includes somatic (muscle and bone) or visceral (organs) pain. Somatic pain is typically localized and described as "sharp", "aching", or "throbbing" whereas visceral nociceptive pain is poorly localized and described in a variety of ways ("gnawing", "cramping", "pressure"). Neuropathic pain is typically due to damage to nerves and is described as "burning", "needles", or "numbness", or is aggravated by touch [62].

Much of the pain experienced in neuro-oncology patients is of mixed type as brain tumors directly press on neuronal tissue and therefore a mixed approach is commonly undertaken.

Nonpharmacologic treatments should accompany the pharmacological approach for patients with any severity of pain; these include environmental changes, treating comorbid symptoms [63] (fear, anxiety, depression), and integrative approaches [64], such as: hypnosis [65], mind-body therapies [66,67], heat and cold stimulation, massage, acupuncture, physical therapy, exercise, biofeedback, art therapy, guided imagery, and distraction.

Pharmacologically, mild nociceptive pain is treated with acetaminophen; moderate to severe nociceptive pain is treated with opioids. For those with pain at least every other day, a long-acting, scheduled agent as well as a short acting as-needed agent should be employed. Short acting agents should be 10–20% of the daily opioid morphine equivalent dose and given every 2–4 h as needed. Methadone is a very useful long acting *mu*-agonist and NMDA antagonist [68] agent available in a variety of forms for easy pediatric dosing. Consultation with a provider comfortable with dosing and management may be necessary [69,70]. Fentanyl patches provide transdermally delivered opioid that can be used in opioid tolerant patients; however, doses are fixed and therefore can be difficult to titrate in smaller patients. In addition, morphine, and oxycodone are available in extended release formulations for use in patients as an oral alternative to methadone and fentanyl patches. These agents come in uncrushable, fixed-dose tablets, which precludes their use in small patients and those that cannot swallow pills. Dosing of extended release (ER) morphine and oxycodone is generally based on a patient's current daily opioid use. For example, a patient taking morphine 5 mg every 4 h around the clock is taking 30 mg per day. This patient would be started on 15 mg morphine ER twice daily. Breakthrough pain doses of 10–20% of daily doses can be offered in addition every 4 h; in this case, 3–6 mg would be appropriate. See Table 3 for nociceptive pain medications and dosages.

In patients with recalcitrant pain, a pain titration may become necessary. The precept of a pain titration is to quickly titrate medication to relief, or dose-limiting side effects. A pain titration is accomplished with interval intravenous (IV) dosing of morphine (or hydromorphone), beginning with an appropriate initial dose for weight and re-assessing response every 30 min. After each assessment using a validated pain assessment scale, additional morphine doses are given as follows: (1) If no relief and pain is >7, administer and increase dose by 50–100%, (2) if partial relief, but pain still >5, give one-half to 1 times the initial dose, and (3) if pain score is 4 or less, then no dose is given at the 30-min assessment. This process is repeated until the patient's pain is 4 or less, or intolerable side effects occur. The total milligrams of opioid required to achieve pain control is calculated by dividing the total morphine dose given by the total number of elapsed hours since initiation of the pain titration, thus providing the hourly morphine rate. With this information, one can calculate an equivalent daily methadone dose, fentanyl patch, or a basal morphine IV/SQ infusion rate. Breakthrough pain treatment with additional doses of morphine are then added at 20% of the daily morphine dose and offered every 3 h IV prn. Alternatively, a patient-controlled analgesia (PCA) pump can be started with a morphine basal rate equivalent to the calculated hourly morphine rate, an added "PCA" or "demand" dose of 100% of the hourly rate (range 50–200%) and a timed "lock-out" of 10 min. Re-assessment of pain should occur multiple times per day; increases in pain and opioid use should be followed by dose adjustments. In the terminal setting, it is not uncommon for 80% of the daily opioid used each day to be delivered continuously via methadone or an IV opioid infusion, and for patients to need a twice daily upward titration [71]. Continuous opioid rate should be adjusted as often as every 6–8 h (time to reach steady state) to attain the goal of demand dosing being required less than twice per hour while awake, and to allow for uninterrupted periods of sleep. Demand and interval breakthrough pain doses can be adjusted after three doses. Additional bolus doses of two times the PCA demand dose can be given every 1–2 h while awaiting therapeutic opioid levels. See Table 4 for initial pediatric PCA dosing.

Table 3. Nociceptive agents.

Drug	Route	Dose	Onset (min)	Peak Effect (h)	Duration (h)	Initial Scheduled Dosing in Opioid Naïve Patients	Available Oral Dose Formulations	Notes
Acetaminophen	Oral, IV, Rectal	10 mg/kg IV q6 hours or 15 mg/kg PO q4 hours.						Avoid in liver disease or consult with hepatologist / GI specialist regarding dosing.
	Initial short-acting dose in an opioid naive patient							
	Route	Dose						
Tramadol	PO	1–2 mg/kg/dose (max initial dose 25–50 mg); Maximum daily dose 400 mg	30–60	1.5	3–7	Short-acting: Every 4–6 hours. Long acting: Every 12 hours	Short-acting: 50 mg tablets Long-acting: 100, 200, 300 mg tablets	Not approved for children less than 18 years of age. May lower seizure threshold. Increased risk of Serotonin Syndrome.
Hydrocodone	PO	0.1–0.2 mg/kg/dose (max 5–10 mg)	10–20	1–3	4–8	Short-acting: Every 6 hours	Short-acting in combination with acetaminophen: 5, 7.5, 10 mg tablets; 2.5 mg/5 mL liquid	Hydrocodone used for pain is only available in combination with acetaminophen or ibuprofen.
Morphine	PO	0.2–0.5 mg/kg/dose (max 5–15 mg)	30	0.5–1	3–6	Short-acting: PO: Every 4 hours. Long-acting: Every 12 hours	Short-acting: 15, 30 mg tablets; 10 mg/5 mL, 20 mg/5 mL, 20 mg/1 mL liquid. Long acting: 15, 30, 60, 100, 200 mg tablets	Short-acting preparation can be compounded into very concentrate SL drops (20 mg/mL). Long acting morphine for opioid tolerant patients only.
	IV/SC	0.05–1 mg/kg/dose (max 2–3 mg)	5–10	N/A	N/A	Every 4 hours	N/A	
Oxycodone	PO	0.1–0.2 mg/kg/dose (max 5–10 mg)	10–15	0.5–1	3–6	Short-acting: Every 4 hours. Long-acting: Every 12 hours	Short-acting: 5, 15, 30 mg tablets; 5 mg/5 mL, 20 mg/mL liquid. Long-acting: 10, 15, 20, 30, 40, 60, 80 mg tablets	Available alone or in combination with acetaminophen. Long acting form for opioid tolerant patients only.
Hydromorphone	PO	0.03–0.06 mg/kg/dose (max 1–3 mg)	15–30	0.5–1	3–5	Short-acting: Every 4 hours; long acting: Once daily	Short-acting: 2, 4, 8 mg tablets; 1 mg/mL liquid Long-acting: 8, 12, 16, 32 mg tablets	Long acting form is for opioid tolerant patients only.
	IV/SC	0.01–0.015 mg/kg/dose (max 0.5–1.5 mg)	15–30	N/A	4–5	Every 4 hours	N/A	
Methadone	PO/SC/PO	0.04 mg/kg/dose BID and titrated weekly to effect	30 min (PO)	3–5 days	Increases with repeater doses up to 60 hours		Tablet, Liquid	Consult expert provider. May prolong QTc; check baseline ECG.

Table 4. Intravenous patient controlled analgesic starting dose recommendations.

Drug	Demand Dose	Lockout Interval	Continuous Rate	4 h Limit
Morphine	0.025 mg/kg (max 2 mg)	10–12 min	0.015 mg/kg/h	0.3–0.4 mg/kg
Hydromorphone	0.005 mg/kg (max 0.3 mg)	6–10 min	0.003 mg/kg/h	0.06–0.08 mg/kg
Fentanyl	0.25 mCg/kg (max 20 mCg)	6–10 min	0.015 mCg/kg/h	3–4 mCg/kg

Advantages of PCA use include: (1) The ability of patients to titrate pain medication to relief; (2) reduced delay in addressing changes in pain intensity; and (3) gives children and young adults some control over their symptom management. Contraindications to the use of PCA are all relative in the end-of-life setting. However, PCA is not an ideal delivery mechanism for patients that are delirious, encephalopathic, or physically unable to press the demand button. If the patient is actively dying, and goals of care are well established, and comfort-directed, then PCA by-proxy (nurse or family member) is appropriate and an often-optimal way to provide pain management [72].

Common opioid side effects include pruritis, nausea, sedation, and constipation. In general, sedation is dose-related and should be avoided, except at end-of-life if consistent with goals. Constipation should be anticipated with all opioids except methadone, and prophylaxis is appropriate in the setting of "around the clock" opioid treatment. Orders for antihistamine are also indicated for as needed use for pruritis. Nausea can be treated with antiemetics. In general, when side effects are not tolerated, opioids can be rotated as patients respond differently to different opioids and, in general, hydromorphone is less likely to cause pruritis than morphine. When rotating opioids, the equivalent opioid dose should be calculated and then reduced by 25% to account for incomplete cross-tolerance. See Table 5 for side effects and suggested treatments.

Table 5. Common or important opioid adverse effects and treatments.

Adverse Effect	Treatments
Constipation	-Polyethylene glycol: 0.5–1.5 GM/kg PO daily. -Senna: 4.3–17.2 mg/day PO (2–6 years), 6–50 mg/day PO (6–12 years), 12–100 mg/day (12 years and older).
Pruritis	-Hydroxyzine (preferred for least sedation): 50 mg/day PO divided every 6 h (<6 years), 50–100 mg/day PO divided every 6 h (≥6 years). -Diphenhydramine: 6.25 mg q4 hours as needed (2–6 years), 12.5–25 mg PO q4 hours as needed (6–12 years), 25–50 mg PO q 4 hours as needed (12 years and older). -Opioid rotation (switch opioids with 25% dose reduction). -Naloxone 0.25–1 microgram/kg/hour IV infusion.
Urinary Retention	-Oxybutinin: 0.2 mg/kg/dose (max 5 mg/dose) PO TID (≤5 years old), 5 mg/dose TID for >5 years old. -Relieve with catheterization, then lower dose or rotate opioid with 25% dose reduction.
Euphoria/Dysphoria	-Lower dose or rotate opioid with 25% dose reduction.
Somnolence	-Lower dose or rotate opioid with 25% dose reduction. -methylphenidate 0.3 mg/kg/dose, (max initial dose 2.5–5 mg/dose) given before breakfast and before lunch (≥6 years old)

Reassessment is critical to pain control and should be performed at a frequency consistent with the medications used (consider expected time-to-onset and duration of action) and pain severity. Children suffering pain deserve aggressive pain control, which may require escalating doses of opioids significantly and rapidly. There is no empiric upper limit dose on opioids; however, an individual may reach their upper limit when he/she experiences increased side effects without additional benefit to pain control [73]. The treatments should follow the "2-step ladder" determined by the World Health Organization: Step 1 includes acetaminophen and/or NSAIDs for mild pain +/− non-opioid adjuvants; and Step 2 adds opioids for more severe pain [74]. Medication choices should be personalized for the patient, considering allergies, comorbidities, drug interactions, adverse effects, social support, and cultural perspectives.

Neuropathic pain in children is treated primarily with gabapentinoids, and/or tricyclic antidepressants, which decrease central nervous system (CNS) excitatory neurotransmission [75]. Again, methadone, with its NMDA antagonism, is also a helpful medication when pain is severe and chronic [48]. Clonidine, topiramate, and duloxetine may also be useful, particularly when agitation, headache, or depressed mood are present, respectively. See Table 6 for neuropathic agents. Adjuvant medications to treat comorbid painful conditions, such as muscle spasms, abdominal cramps, and chemotherapy induced peripheral neuropathy, are shown in Table 7.

Though beyond the scope of this text, pain management can also include palliative radiation and surgical approaches, such as dorsal rhizotomy. Consultation with radiation oncology, neurosurgery, pain medicine, and/or hospice and palliative medicine specialists is prudent in patients with recalcitrant pain. In the rare case where pain control cannot be achieved, the goals-of-care may be consistent with palliative sedation, which necessitates consultation with palliative care specialists [71,76–78].

Table 6. Neuropathic agents.

Drug	Dose	Notes
Gabapentin	Day 1: 5 mg/kg/dose (max 300 mg/dose) PO at bedtime. Day 2: 5 mg/kg/dose (max 300 mg/dose) PO BID. Day 3: 5 mg/kg/dose (max 300 mg/dose) PO TID. Dose may be further titrated to a maximum dose of 50 mg/day (and generally no more than 1800 mg/day).	Comes in a liquid. May cause drowsiness, dizziness, and peripheral edema. Dose adjust for renal impairment.
Pregabalin	75 mg BID.	Initial adult dose; can titrate up to 300 BID max.
Clonidine	Oral: Immediate release: Initial: 2 mCg/kg/dose every 4 to 6 h; increase incrementally over several days; range: 2 to 4 mCg/kg/dose every 4 to 6 h. Topical: Transdermal patch: May be switched to the transdermal delivery system after oral therapy is titrated to an optimal and stable dose; a transdermal dose is approximately equivalent to $\frac{1}{2}$ to 1 × the total oral daily dose.	Limited data available for pain in children and adolescents. Helps with opioid withdrawal, helps with sleep. Can lower BP. Good for dysautonomia pain.
Topiramate	-6–12 years (weight greater than or equal to 20 kg): 15 mg PO daily for 7 days, then 15 mg PO BID. -≥12 years: 25 mg PO at bedtime for 7 days, then 25 mg PO BID and titrate up to 50 mg PO BID. -Maximum daily dose 200 mg.	May cause acidosis, drowsiness, dizziness, and nausea. Dose adjust for renal impairment and hepatic dysfunction.
Amitriptyline	-0.1 mg/kg PO at bedtime. -titrate as tolerated over 3 weeks to 0.5–2 mg/kg at bedtime. -Maximum: 25 mg/dose.	Consider for continuous and shooting neuropathic pain. Caution use in patients with arrhythmias. May cause sedation, arrhythmias, dry mouth, orthostasis, and urinary retention. Caution use in patients with seizures; avoid MAOIs, other SSRIs, or SNRIs due to potential for serotonin syndrome.
Duloxetine	Approved for anxiety in children >7 years. Start with 30 mg capsule at bedtime and can titrate up to 60 mg qHS.	Antidepressants can increase suicidal thinking in pediatric patients with major depressive disorder. Duloxetine may increase the risk of bleeding events. Concomitant use of aspirin, nonsteroidal anti-inflammatory drugs, warfarin, and other anti-coagulants may add to this risk. Taper slowly.

Table 7. Adjuvant pain treatments.

Drug	Indication	Dose	Notes
Dexamethasone	Inflammation, Nerve compression	-1 mg/kg/day IV or PO in divided doses every 6 h). -Maximum: 16 mg/day. Use lowest effective dose.	May cause impaired healing, infection, thrush, hyperglycemia, weight gain, myopathy, stomach upset, psychosis, emotional instability.
Diazepam	Muscle spasms	Oral: Children: 0.12 to 0.8 mg/kg/day in divided doses every 6 to 8 h.	
Tizanidine	Muscle spasms	Children 2 to <10 years: Oral: 1 mg at bedtime, titrate as needed. Children ≥10 years and Adolescents: Oral: 2 mg at bedtime, titrate as needed.	Oral: Titrate initial dose upward to reported effective range of: 0.3 to 0.5 mg/kg/day in 3 to 4 divided doses; maximum daily dose: 24 mg/day.
Cyclobenzaprine	Muscle spasms	Greater than or equal to 15 years old: 5 mg PO three times daily Maximum 30 mg/day.	
Dicyclomine	Abdominal cramping	Infants ≥6 months and Children <2 years: Oral: 5 to 10 mg 3 to 4 times daily administered 15 min before feeding. Children ≥2 years Oral: 10 mg 3 to 4 times daily Adolescents: Oral: 10 to 20 mg 3 to 4 times daily. If efficacy not achieved in 2 weeks, therapy should be discontinued.	
5% lidocaine patch	Nociceptive or neuropathic pain	1–3 patches applied daily (depending on size) up to 12 h per day.	Can be cut to fit.
OTC creams	Nociceptive or neuropathic pain	Apply topically to localized areas of neuropathic pain BID-TID.	
Prescription creams: Diclofenac cream; compounded neuropathic agents	Nociceptive or neuropathic pain	Apply topically to localized areas of neuropathic pain BID-TID.	
Ice, heat	Nociceptive or neuropathic pain		

9. Altered Mood

As brain tumors progress, they often (in 60% of children) cause mood changes, such as depression or anxiety, which are important [79], yet under-addressed, forms of suffering. In the setting of life-limiting illness, such mood disorders are likely biochemical in nature as well as "supratentorial" or cognitive in nature. Therefore, appropriate care assesses and treats both etiologies. Adolescents, in particular, demonstrate accelerated transitions, notably an "illness transition" in which one identifies with the illness as being part of their being, and a "developmental transition" in which development behaviors change in response to the disease [80]. However, their growth comes with feelings of sadness, anxiety, difficulty speaking with their parents, and fear of being alone; and unlike pain, which is addressed more than 80% of the time, these emotional symptoms are far less frequently addressed (around 45% of the time) [81]. Depressed mood or anxiety should be treated based on severity at presentation. For all patients with mood changes and unmanaged stress (e.g., fear, changes in function, uncontrolled symptoms), integrative approaches and non-pharmacological options including cognitive behavioral therapy, family therapy, and massage/relaxation therapy can offer benefit. Music, along with other forms of art, are increasingly used to help patients and family members express their emotions [82–84]. For patients presenting with mild symptoms rooted primarily in an uncertainty of the future and fear of death, dying support through attention, education, and expressive supportive counseling is appropriate. Therapeutic counseling is provided by a variety of clinicians, using a myriad of techniques, including: Meeting the patients where they are in development, allowing time and presence to validate emotions, and compassionate, honest discussion of the anticipated disease trajectory. For example, family pastors, hospital chaplains, and bedside nurses (among others) can assess the child's belief system and address spiritual/existential fears, as well as engage in reflective listening, prayer, touch, silence, and ritual [85–87]. Social workers play a critical role in helping patients achieve their personal goals of legacy and in leveraging the patient's support network for direct support to the patient [88,89]. Psychologists can assess the patient for specific fears and provide cognitive behavioral counseling [90].

For moderate symptom burden, cognitive behavior therapy in addition to medications, such as selective serotonin reuptake inhibitors (SSRIs), selective norepinephrine reuptake inhibitors (SNRIs), and tricyclic antidepressants, should be initiated and selected primarily based on the secondary beneficial effects (i.e., SNRIs for neuropathic pain component, SSRIs for weight gain, etc.). For severe mood symptoms, medications and supportive therapies should be initiated together, psychology and/or psychiatry consultation should be obtained, and disposition should be considered (i.e., admission to facility for safety if suicidal).

Children and adolescents laden with symptoms and the weight of disease, may express their burden through requests for a hastening of death. American Academy of Pediatrics (AAP) guidelines do not support physician-assisted-death; the request should be met with normalization, compassionate care, and an aggressive focus on relieving burdens and increasing quality of life [91–93]. Psychological consultation is important. It is important to remember and communicate that foregoing interventions and thus allowing a disease to progress along its natural course to death is ethically sound, is not suicide, and is not physician-assisted death [94,95] Elucidating the patient's voice about their hopes and fears is powerful; Voicing My Choices [96] and Five Wishes [97] are two studied, standardized tools [98].

10. Agitation, and Altered Mental Status

Many patients with brain tumors demonstrate agitation, or delirium [35,41,47].

Agitation can be caused by direct tumor effects, medical complications (e.g., uremia), and medications (e.g., opioid). Benzodiazepines (midazolam, lorazepam, diazepam, clonazepam) at the lowest possible dose are the primary treatment, especially if anxiety is present, followed by neuroleptics (haloperidol, thioridazine, chlorpromazine, risperidone), and alpha-agonists (clonidine) [49]. Neuroleptics are particularly helpful when delirium is present. See Table 8 for dosing recommendations.

Table 8. Medical treatments for agitation.

Drug	Route	Dose	Available Oral Dose Formulations
Lorazepam	PO, IV	0.05 mg/kg/dose PO/IV q4 hours; max single dose 2 mg.	Tablet: 0.5, 1, 2 mg. Oral solution 2 mg/mL.
Diazepam	PO, IV, IM	0.12–0.8 mg/kg/day PO divided q6 hours. 0.04–0.2 mg/kg IV/IM q2 hours, max. 0.6 mg/8 h.	Tablet: 2, 5, 10 mg. Solution: 1 mg/mL, 5 mg/mL.
Clonazepam >6 years	PO, IV	<30 kg: 0.01–0.03 mg/kg/day PO divided q8 hours; increase by 0.25–0.5 mg/day q3 days; maximum 0.1–0.2 mg/kg/day PO divided q8 hours. >30 kg: 1.5 mg/day PO divided q8 hours; increase by 0.5–1 mg q3 days; maximum 20 mg/day.	Tablet
Midazolam	PO, IV, intranasal	500–750 mCg/kg PO once prior to procedure.	Oral solution: 2 mg/mL.
Haloperidol ≥3 years	PO, IM	Oral: 0.01 mg/kg/dose 3 times daily as needed. Starting dose max 0.5 mg/day. Titrate slowly as directed.	Tablet: 0.5, 1, 2, 5, 10, 20 mg. Oral solution: 2 mg/mL.
Chlorpromazine ≥6 months	PO, IV	Initial: 0.55 mg/kg/dose PO every 4–6 h as needed. Common initial dose 10–25 mg. Max daily dose ≤5 years old: 50 mg/day; >5 years old 200 mg/day.	Tablet: 10, 25, 50, 100 mg.
Risperidone	PO	>5 years old and 15–20 kg: 0.25 mg/day qHS; >20 kg 0.5 mg/day qHS or divided BID. Can titrate up 100% after 4 days.	Tablet (and orally dissolving tablet): 0.25, 0.5, 1, 2, 3, 4 mg. Oral solution: 1 mg/mL.
Clonidine	PO, IV, transdermal	0.1–0.5 mg PO q8 hours; titrate up slowly every 3 days; wean upon discontinuing.	Tablet: 0.1, 0.2, 0.3 mg. Extended release: 0.1 mg. Patch: 0.1, 0.2, 0.3 mg/day.

An altered mental status is the most common symptom as children with tumors of the brain approach death (75% of children within the last week of life; 85% overall) [33,47]. This usually manifests as a slow decrease in consciousness over days to weeks. While treatment options are few at this stage of disease, it is important that clinicians provide anticipatory guidance to family and other providers.

11. Care of the Family, Caregiver, and Siblings

Palliative care teams maintain the goal of caring not only for the patient, but also for the entire family as they grieve and bereave [99]. The complicated emotion and process of grief and bereavement, respectively, begin at the time of diagnosis and accelerate as attention is turned from cure of disease to palliation of symptoms. Finding meaning and realistic hoped-for goals can soothe patients and families in this phase of care.

While oncology, palliative care, and other interdisciplinary teams have matured areas of disease and symptom management, families report significant deficiencies in psychological care near the end of life [100]. Many family members report unaddressed psychospiritual distress and significant caregiver burdens [93]. Consultation with physical therapy, occupational therapists, child life specialists, art therapists, chaplains, and hospices can provide additional supportive resources. Siblings deserve particular attention and developmentally appropriate interventions, such as cognitive behavioral therapy, art, and play therapies, and close bereavement follow-up.

Similarly, regardless of the prognosis or the choices in care that parents select, disease-specific or pediatric oncological support communities can provide a community, validation, and education to the entire family unit; providers should try to connect families to such a group whenever possible.

12. Summary

Pediatric patients suffering from CNS tumors face difficult clinical decisions and carry a significant symptom burden. Given the high risk of mortality, communication of prognosis and goal-setting can be challenging. Supporting caregiver decision making requires honest, empathic communication and attention to the family values and emotions. Ongoing symptom assessment and management is a must to optimize quality of life and reduce suffering. Providers caring for these children should have a basic knowledge of palliative communication skills and pain and symptom management, with access to specialist palliative care providers as needed.

Author Contributions: Each author contributed substantially to the writing and revisions: Writing—original draft preparation P.H.B.; Writing—Review and Editing K.M.

Funding: This research received no external funding.

Conflicts of Interest: The authors declare no conflict of interest.

References

1. Ostrom Quinn CBTRUS Statistical report: Primary brain and other central nervous system tumors diagnosed in the US 2010–2014. *Neuro-Oncology* **2017**, *19*, 1–88. [CrossRef] [PubMed]
2. Ward, E.; DeSantis, C.; Robbins, A.; Kohler, B.; Jemal, A. Childhood and adolescent cancer statistics, 2014. *CA Cancer J. Clin.* **2014**, *64*, 83–103. [CrossRef] [PubMed]
3. *Committee on Approaching Death: Addressing Key End of Life Issues; Institute of Medicine. Dying in America: Improving Quality and Honoring Individual Preferences near the End of Life*; National Academies Press (US): Washington, DC, USA, 2015.
4. Approaching Death: Improving Care at the End of Life—A Report of the Institute of Medicine. *Health Serv. Res.* **1998**, *33*, 1–3.
5. Pediatric Palliative Care and Hospice Care Commitments, Guidelines, and Recommendations. Section on Hospice and Palliative Medicine and committee on Hospital Care. *Pediatrics* **2013**, *132*, 966–972. [CrossRef]
6. World Health Organization. *Cancer Pain Relief and Palliative Care in Children*; World Health Organization: Geneva, Switzerland, 1998.

7. Wolfe, J. Suffering in children at the end of life: Recognizing an ethical duty to palliate. *J. Clin. Ethics* **2000**, *11*, 157–163. [PubMed]

8. Fischer, C.; Petriccione, M.; Donzelli, M.; Pottenger, E. Improving Care in Pediatric Neuro-oncology Patients: An Overview of the Unique Needs of Children with Brain Tumors. *J. Child Neurol.* **2016**, *31*, 488–505. [CrossRef] [PubMed]

9. Vargo, M. Brain tumor rehabilitation. *Am. J. Phys. Med.* **2011**, *90*, S50–S62. [CrossRef] [PubMed]

10. Abdel-Baki, M.S.; Hanzlik, E.; Kieran, M.W. Multidisciplinary pediatric brain tumor clinics: The key to successful treatment? *CNS Oncol.* **2015**, *4*, 147–155. [CrossRef] [PubMed]

11. Hill, D.L.; Nathanson, P.G.; Carrol, K.W.; Schall, T.E.; Miller, V.A.; Feudtner, C. Changes in Parental Hopes for Seriously Ill Children. *Pediatrics* **2018**, *141*, e20173549. [CrossRef] [PubMed]

12. Rosenberg, A.R.; Feudtner, C. What else are you hoping for? Fostering hope in paediatric serious illness. *Acta Paediatr.* **2016**, *105*, 1004–1005. [CrossRef] [PubMed]

13. Marston, J.M. The Spirit of "Ubuntu" in Children's Palliative Care. *J. Pain Sympt. Manag.* **2015**, *50*, 424–427. [CrossRef] [PubMed]

14. Hendricks-Ferguson, V.L.; Pradhan, K.; Shih, C.S.; Gauvain, K.M.; Kane, J.R.; Liu, J.; Haase, J.E. Pilot Evaluation of a Palliative and End-of-Life Communication Intervention for Parents of Children with a Brain Tumor. *J. Pediatr. Oncol. Nurs.* **2017**, *34*, 203–213. [CrossRef] [PubMed]

15. Mack, J.W.; Fasciano, K.M.; Block, S.D. Communication about Prognosis with Adolescent and Young Adult Patients with Cancer: Information Needs, Prognostic Awareness, and Outcomes of Disclosure. *J. Clin. Oncol.* **2018**, *36*, 1861–1867. [CrossRef] [PubMed]

16. Singer, A.E.; Ash, T.; Ochotorena, C.; Lorenz, K.A.; Chong, K.; Shreve, S.T.; Ahluwalia, S.C. A Systematic Review of Family Meeting Tools in Palliative and Intensive Care Settings. *Am. J. Hosp. Palliat. Care* **2016**, *33*, 797–806. [CrossRef] [PubMed]

17. Baile, W.F.; Buckman, R.; Lenzi, R.; Glober, G.; Beale, E.A.; Kudelka, A.P. SPIKES-A six-step protocol for delivering bad news: Application to the patient with cancer. *Oncologist* **2000**, *5*, 302–311. [CrossRef] [PubMed]

18. Smith, R.C.; Hoppe, R.B. The patient's story: Integrating the patient- and physician-centered approaches to interviewing. *Ann. Intern. Med.* **1991**, *115*, 470–477. [CrossRef] [PubMed]

19. Back, A.L.; Anderson, W.G.; Bunch, L.; Marr, L.A.; Wallace, J.A.; Yang, H.B.; Arnold, R.M. Communication about cancer near the end of life. *Cancer* **2008**, *113*, 1897–1910. [CrossRef] [PubMed]

20. Egan, G. *The Skilled Helper: A Problem-Management and Opportunity-Development Approach to Helping*; Thomson Learning: Boston, MA, USA, 2002; Volume 8.

21. Levetown, M.; Carter, M.A. Child-centered care in terminal illness: An ethical framework. In *Oxford Textbook of Palliative Medicine*, 2nd ed.; Doyle, D., Hanks, G.W.C., MacDonald, N., Eds.; Oxford University Press: New York, NY, USA, 1998; pp. 1107–1117.

22. Bluebond-Langner, M. *The Private Worlds of Dying Children*; Princeton University Press: Princeton, NJ, USA, 1978.

23. Miller, E.G.; Feudtner, C. Health reform law allows children in hospice to be treated for their disease. *AAP News* **2013**, *34*, 14. [CrossRef]

24. Levine, D.R.; Johnson, L.M.; Mandrell, B.N.; Yang, J.; West, N.K.; Hinds, P.S.; Baker, J.N. Does phase 1 trial enrollment preclude quality end-of-life care? Phase 1 trial enrollment and end-of-life care characteristics in children with cancer. *Cancer* **2015**, *121*, 1508–1512. [CrossRef] [PubMed]

25. Chris, F. Collaborative Communication in Pediatric Palliative Care: A Foundation for Problem-Solving and Decision-Making. *Pediatr. Clin. N. Am.* **2007**, *54*, 583–607.

26. Anthony, B.; Robert, A. Dealing with Conflict in Caring for the Seriously Ill 'It Was Just Out of the Question'. *JAMA* **2005**, *293*, 1374–1381.

27. Brian, C.; Marcia, L.; Sarah, F. *Palliative Care for Infants, Children, and Adolescents: A Practical Handbook*; JHU Press: Baltimore, MD, USA, 2013.

28. Kuhlen, M.; Hoell, J.; Balzer, S.; Borkhardt, A.; Janssen, G. Symptoms and management of pediatric patients with incurable brain tumors in palliative home care. *Eur. J. Paediatr. Neurol.* **2016**, *20*, 261–269. [CrossRef] [PubMed]

29. Veldhuijzen van Zanten, S.E.; van Meerwijk, C.L.; Jansen, M.H.; Twisk, J.W.; Anderson, A.K.; Coombes, L.; Breen, M.; Hargrave, O.J.; Hemsley, J.; Craig, F.; et al. Palliative and end-of-life care for children with diffuse

intrinsic pontine glioma: Results from a London cohort study and international survey. *Neuro-Oncology* **2016**, *18*, 582–588. [CrossRef] [PubMed]

30. Pritchard, M.; Burghen, E.; Srivastava, D.K.; Okuma, J.; Anderson, L.; Powell, B.; Furman, W.L.; Hinds, P.S. Cancer-related symptoms most concerning to parents during the last week and last day of their child's life. *Pediatrics* **2008**, *121*, e1301–e1309. [CrossRef] [PubMed]
31. Arland, L.C.; Hendricks-Ferguson, V.L.; Pearson, J.; Foreman, N.K.; Madden, J.R. Development of an in-home standardized end-of-life treatment program for pediatric patients dying of brain tumors. *JSPN* **2013**, *18*, 144–157. [CrossRef] [PubMed]
32. Vallero, S.G.; Lijoi, S.; Bertin, D.; Pittana, L.S.; Bellini, S.; Rossi, F.; Peretta, P.; Basso, M.E.; Fagioli, F. End-of-life care in pediatric neuro-oncology. *Pediatr. Blood Cancer* **2014**, *61*, 2004–2011. [CrossRef] [PubMed]
33. Protus, B.M.; Brauer, P.A.; Kimbrel, J.M. Evaluation of Atropine 1% Ophthalmic Solution Administered Sublingually for the Management of Terminal Respiratory Secretions. *Am. J. Hosp. Palliat. Care* **2013**, *30*, 388–392. [CrossRef] [PubMed]
34. Diekeman, D.S.; Botkin, J.R. Clinical Report-Forgoing Medically Provided Nutrition and Hydration in Children. *Pediatrics* **2009**, *124*, 813–822. [CrossRef] [PubMed]
35. Zelcer, S.; Cataudella, D.; Cairney, A.E.; Bannister, S.L. Palliative care of children with brain tumors: A parental perspective. *Arch. Pediatr. Adolesc. Med.* **2010**, *164*, 225–230. [CrossRef] [PubMed]
36. Cataudella, D.A.; Zelcer, S. Psychological experiences of children with brain tumors at end of life: Parental perspectives. *J. Palliat. Med.* **2012**, *15*, 1191–1197. [CrossRef] [PubMed]
37. Fager, S.; Bardach, L.; Russell, S.; Higginbotham, J. Access to augmentative and alternative communication: New technologies and clinical decision-making. *J. Pediatr. Rehabil. Med.* **2012**, *5*, 53–61. [PubMed]
38. Korner, S.; Sieniawski, M.; Kollewe, K.; Rath, K.J.; Krampfl, K.; Zapf, A.; Dengler, R.; Petri, S. Speech therapy and communication device: Impact on quality of life and mood in patients with amyotrophic lateral sclerosis. *Amyotroph. Lateral. Scler. Frontotemporal. Degener.* **2013**, *14*, 20–25. [CrossRef] [PubMed]
39. Russell, C.E.; Bouffet, E. Profound answers to simple questions. *J. Clin. Oncol.* **2015**, *33*, 1294–1296. [CrossRef] [PubMed]
40. Pace, A.; Parisi, C.; Di Lelio, M.; Zizzari, A.; Petreri, G.; Giovannelli, M.; Pompili, A. Home rehabilitation for brain tumor patients. *J. Exp. Clin. Cancer Res.* **2007**, *26*, 297–300. [PubMed]
41. Pace, A.; Di Lorenzo, C.; Guariglia, L.; Jandolo, B.; Carapella, C.M.; Pompili, A. End of life issues in brain tumor patients. *J. Neuro-Oncol.* **2009**, *91*, 39–43. [CrossRef] [PubMed]
42. Madden, K.; Bruera, E. Very-Low-Dose Methadone to Treat Refractory Neuropathic Pain in Children with Cancer. *J. Palliat. Med.* **2017**, *20*, 1280–1283. [CrossRef] [PubMed]
43. Wusthoff, C.J.; Shellhaas, R.A.; Licht, D.J. Management of common neurologic symptoms in pediatric palliative care: Seizures, agitation, and spasticity. *Pediatr. Clin. N. Am.* **2007**, *54*, 709–733. [CrossRef] [PubMed]
44. Pace, A.; Villani, V.; Di Lorenzo, C.; Guariglia, L.; Maschio, M.; Pompili, A.; Carapella, C.M. Epilepsy in the end-of-life phase in patients with high-grade gliomas. *J. Neuro-Oncol.* **2013**, *111*, 83–86. [CrossRef] [PubMed]
45. Krouwer, H.G.; Pallagi, J.L.; Graves, N.M. Management of seizures in brain tumor patients at the end of life. *J. Palliat. Med.* **2000**, *3*, 465–475. [CrossRef] [PubMed]
46. Ekokobe, F.; Bricke, P.; Mungall, D.; Aceves, J.; Ebwe, E.; Tang, W.; Kirmani, B. The Role of Levetiracetam in Treatment of Seizures in Brain Tumor Patients. *Front. Neurol.* **2013**, *4*, 153.
47. Major, P.; Greenberg, E.; Khan, A.; Thiele, E.A. Pyridoxine supplementation for the treatment of levetiracetam-induced behavior side effects in children: Preliminary results. *Epilepsy Behav.* **2008**, *13*, 557–559. [CrossRef] [PubMed]
48. Santucci, G.; Mack, J.W. Common Gastrointestinal Symptoms in Pediatric Palliative Care: Nausea, Vomiting, Constipation, Anorexia, Cachexia. *Pediatr. Clin. N. Am.* **2007**, *54*, 673–689. [CrossRef] [PubMed]
49. Adult Guidelines for Assessment and Management of Nausea and Vomiting. Dana Farber Cancer Institute. Available online: http://pinkbook.dfci.org/ (accessed on 15 July 2018).
50. Pediatric Palliative Care. Available online: https://www.uptodate.com/contents/pediatric-palliative-care (accessed on 15 July 2018).
51. Children's Oncology Group. Guideline for the Prevention of Nausea and Vomiting Due to Antineoplastic Medication in Pediatric Cancer Patients. Version September 2015. Available online: https://childrensoncologygroup.org/downloads/COG_SC_CINV_Guideline_Document.pdf (accessed on 15 July 2018).

52. Treatment-Related Nausea and Vomiting (PDQ®)—Health Professional Version. Available online: https:// www.cancer.gov/about-cancer/treatment/side-effects/nausea/nausea-hp-pdq (accessed on 15 July 2018).

53. Jacknow, D.S.; Tschann, J.M.; Link, M.P.; Boyce, W.T. Hypnosis in the prevention of chemotherapy-related nausea and vomiting in children: A prospective study. *J. Dev. Behav. Pediatr.* **1994**, *15*, 258–264. [CrossRef] [PubMed]

54. Cotanch, P.; Hockenberry, M.; Herman, S. Self-hypnosis as antiemetic therapy in children receiving chemotherapy. *Oncol. Nurs. Forum* **1985**, *12*, 41–46. [PubMed]

55. Vickers, A.J. Can acupuncture have specific effects on health? A systematic review of acupuncture antiemesis trials. *J. R. Soc. Med.* **1996**, *6*, 303–311. [CrossRef]

56. Zeltzer, L.; LeBaron, S.; Zeltzer, P.M. The effectiveness of behavioral intervention for reducing nausea and vomiting in children and adolescents receiving chemotherapy. *J. Clin. Oncol.* **1984**, *2*, 683–690. [CrossRef] [PubMed]

57. Olson, K.; Amari, A. Self-reported Pain in Adolescents with Leukemia or a Brain Tumor: A Systematic Review. *Cancer Nurs.* **2015**, *38*, E43–E53. [CrossRef] [PubMed]

58. Clark, D. 'Total pain', disciplinary power and the body in the work of Cicely Saunders 1958–1967. *Soc. Sci. Med.* **1999**, *49*, 727–736. [CrossRef]

59. Anita, M.; Chan, L.S. Understanding of the concept of "total pain": A prerequisite for pain control. *J. Hosp. Palliat. Nurs.* **2008**, *10*, 26–32.

60. Manworren, R.C.; Hynan, L.S. Clinical validation of FLACC: Preverbal patient pain scale. *Pediatr Nurs.* **2003**, *29*, 140–146. [PubMed]

61. Whaley, L.; Wong, D.L. *Nursing Care of Infants and Children*, 3rd ed.; The C.V. Mosby Company: St. Louis, MO, USA, 1987.

62. Komatz, K.; Carter, B. Pain and Symptom Management in Pediatric Palliative Care. *Pediatr. Rev.* **2015**, *36*, 527–534. [CrossRef] [PubMed]

63. Hauer, J.; Duncan, J.; Scullion, B.F. Pediatric Pain and Symptom Management Guidelines. Dana Farber Cancer Institute/Boston Children's Hospital Pediatric Advanced Care Team. 2014. Available online: http://pinkbook.dfci.org/assets/docs/blueBook.pdf (accessed on 15 July 2018).

64. Zeltzer, L.K.; Tsao, J.C.; Stelling, C.; Powers, M.; Levy, S.; Waterhouse, M. A phase I study on the feasibility and acceptability of an acupuncture/hypnosis intervention for chronic pediatric pain. *J. Pain Symptom Manag.* **2002**, *24*, 437. [CrossRef]

65. Tsao, J.C.; Meldrum, M.; Kim, S.C.; Jacob, M.C.; Zeltzer, L.K. Treatment Preferences for CAM in children with chronic pain. *Evid. Based Complement. Alternat. Med.* **2007**, *4*, 367–374. [CrossRef] [PubMed]

66. Tsao, J.C.; Zeltzer, L.K. Complementary and Alternative Medicine Approaches for Pediatric Pain: A Review of the State-of-the-science. *Evid. Based Complement. Alternat. Med.* **2005**, *2*, 149–159. [CrossRef] [PubMed]

67. Ott, M.J. Mindfulness meditation in pediatric clinical practice. *Pediatr. Nurs.* **2002**, *28*, 487–490. [PubMed]

68. Madden, K.; Mills, S.; Dibaj, S.; Williams, J.L.; Liu, D.; Bruera, E. Methadone as the Initial Long-Acting Opioid in Children with Advanced Cancer. *J. Palliat. Med.* **2018**. [CrossRef] [PubMed]

69. Madden, K.; Park, M.; Liu, D.; Bruera, E. The frequency of QTc prolongation among pediatric and young adult patients receiving methadone for cancer pain. *Pediatr. Blood Cancer* **2017**, *64*. [CrossRef] [PubMed]

70. Madden, K.; Park, M.; Liu, D.; Bruera, E. Practices, Attitudes, and Beliefs of Palliative Care Physicians Regarding the Use of Methadone and Other Long-Acting Opioids in Children with Advanced Cancer. *J. Palliat. Med.* **2018**. [CrossRef] [PubMed]

71. Gregoire, M.C.; FRager, G. Ensuring pain relief at the end of life. *Pain Res. Manag.* **2006**, *11*, 163–171. [CrossRef] [PubMed]

72. Zernikow, B.; Michel, E.; Craig, F.; Anderson, B.J. Pediatric palliative care: Use of opioids for the management of pain. *Pediatr. Drugs* **2009**, *11*, 129–151. [CrossRef] [PubMed]

73. Hutton, N.; Levetown, M.; Frager, G. The Hospice and Palliative Medicine Approach to Caring for Pediatric Patients. Hospice and Palliative Care Training for Physicians: A Self-Study Program. Available online: https: //www.hopkinsmedicine.org/som/faculty/appointments/ppc/documents/portfolios/Hutton/Hutton-Portfolio-Samples/the-hospice-and-palliative-medicine-approach-to-caring-for-pediatrics-patients.pdf (accessed on 15 July 2018).

74. World Health Organization (WHO). *WHO Guidelines on the Pharmacological Treatment of Persisting Pain in Children with Medical Illnesses*; WHO/IASP: Geneva, Switzerland, 2012.

75. Friedrichsdorf, S.J.; Kang, T.I. The Management of Pain in Children with Life-limiting Illnesses. *Pediatr. Clin. N. Am.* **2007**, *54*, 645–672. [CrossRef] [PubMed]
76. Postovsky, S.; Moaed, B.; Krivoy, E.; Ofir, R.; Ben Arush, M.W. Practice of palliative sedation in children with brain tumors and sarcomas at the end of life. *Pediatr. Hematol. Oncol.* **2007**, *24*, 409–415. [CrossRef] [PubMed]
77. Gibbons, K.; DeMonbrun, A.; Beckman, E.J.; Keefer, P.; Wagner, D.; Stewart, M.; Saul, D.; Hakel, S.; Liu, M.; Niedner, M. Continuous Lidocaine Infusions to Manage Opioid-Refractory Pain in a Series of Cancer Patients in a Pediatric Hospital. *Pediatr. Blood Cancer* **2016**, *63*, 1168–1174. [CrossRef] [PubMed]
78. Johnson, L.M.; Frader, J.; Wolfe, J.; Baker, J.N.; Anghelescu, D.L.; Lantos, J.D. Palliative Sedation with Propofol for an Adolescent with a DNR Order. *Pediatrics* **2017**, *142*, e2017-0487. [CrossRef] [PubMed]
79. Weaver, M.S.; Heinze, K.E.; Bell, C.J.; Wiener, L.; Garee, A.M.; Kelly, K.P.; Casey, R.L.; Watson, A.; Hinds, P.S.; Pediatric Palliative Care Special Interest Group at Children's National Health System. Establishing psychosocial palliative care standards for children and adolescents with cancer and their families: An integrative review. *Palliat. Med.* **2016**, *30*, 212–223. [CrossRef] [PubMed]
80. Johnston, B.; Jindal-Snape, D.; Pringle, J. Life transitions of adolescents and young adults with life-limiting conditions. *Int. J. Palliat. Nurs.* **2016**, *22*, 608–617. [CrossRef] [PubMed]
81. Theunissen, J.M.; Hoogerbrugge, P.M.; van Achterberg, T.; Prins, J.B.; Vernooij-Dassen, M.J.; van den Ende, C.H. Symptoms in the palliative phase of children with cancer. *Pediatr. Blood Cancer* **2007**, *49*, 160–165. [CrossRef] [PubMed]
82. Ready, T. Music as language. *Am. J. Hospice Palliat. Med.* **2010**, *27*, 7–15. [CrossRef] [PubMed]
83. Fagen, T.S. Music therapy in the treatment of anxiety and fear in terminal pediatric patients. *Music Ther.* **1982**, *2*, 13–23. [CrossRef]
84. Stevens, M.M. Psychological adaptation of the dying child. In *Oxford Textbook of Palliative Medicine*, 2nd ed.; Doyle, D., Hanke, G.W.C., MacDonald, N., Eds.; Oxford University Press: New York, NY, USA, 1998; pp. 1045–1056.
85. Michelle, N. Addressing Children's Beliefs through Fowler's Stages of Faith. *J. Pediatr. Nurs.* **2011**, *26*, 44–50.
86. Puchalski, C.; Ferrell, B.; Virani, R.; Otis-Green, S.; Baird, P.; Bull, J.; Chochinov, H.; Handzo, G.; Nelson-Becker, H.; Prince-Paul, M.; et al. Improving the Quality of Spiritual Care as a Dimension of Palliative Care: The Report of the Consensus Conference. *J. Palliat. Med.* **2009**, *12*, 885–904. [CrossRef] [PubMed]
87. Himelstein, B.P. Palliative Care for Infants, Children, Adolescents, and Their Families. *J. Palliat. Med.* **2006**, *9*, 163–181. [CrossRef] [PubMed]
88. Freyer, D.R.; Kuperberg, A.; Sterken, D.J.; Pastyrnak, S.L.; Hudson, D.; Richards, T. Multidisciplinary Care of the Dying Adolescent. *Child Adolesc. Psychiatr. Clin. N. Am.* **2006**, *15*, 693–715. [CrossRef] [PubMed]
89. Michelson, K.; Steinhorn, D. Pediatric End-of-Life Issues and Palliative Care. *Clin. Ped. Emerg. Med.* **2007**, *8*, 212–219. [CrossRef] [PubMed]
90. McSherry, M.; Kehoe, K.; Carroll, J.; Kang, T.; Rourke, M. Psychosocial and Spiritual Needs of Children Living with a Life-Limiting Illness. *Pediatr. Clin. N. Am.* **2007**, *54*, 609–629. [CrossRef] [PubMed]
91. Dyck, A.J. An Alternative to the Ethic of Euthanasia. In *Ethics in Medicine: Historical Perspectives and Contemporary Concerns*; MIT Press: Cambridge, MA, USA, 1977; pp. 529–535.
92. Roy, D.J. On the ethics of euthanasia discourse. *J. Palliat. Care* **1996**, *12*, 3–5. [PubMed]
93. Kenny, N.P.; Frager, G. Refractory symptoms and terminal sedation of children: Ethical issues and practical management. *J. Palliat. Care* **1996**, *12*, 40–45. [PubMed]
94. Billings, J.A.; Block, S.D. Slow euthanasia. *J. Palliat. Care* **1996**, *12*, 21–30. [PubMed]
95. Mount, B. Morphine drips, terminal sedation, and slow euthanasia: Definitions and facts, not anecdotes. *J. Palliat. Care* **1996**, *12*, 31–37. [PubMed]
96. Wiener, L.; Zadeh, S.; Battles, H.; Baird, K.; Ballard, E.; Osherow, J.; Pao, M. Allowing Adolescents and Young Adults to Plan Their End-of-Life Care. *Pediatrics* **2012**, *130*, 897–905. [CrossRef] [PubMed]
97. Wiener, L.; Ballard, E.; Brennan, T.; Battles, H.; Martinez, P.; Pao, M. How I Wish to be Remembered: The Use of an Advance Care Planning Document in Adolescent and Young Adult Populations. *J. Palliat. Med.* **2008**, *11*, 1309–1313. [CrossRef] [PubMed]
98. Zadeh, S.; Pao, M.; Wiener, L. Opening End-of-life discussions: How to introduce Voicing My CHOICES, and advance care planning guide for adolescents and young adults. *Palliat. Support. Care* **2015**, *13*, 591–599. [CrossRef] [PubMed]

99. Stevens, M.M. Care of the dying child and adolescent: Family adjustment and support. In *Oxford Textbook of Palliative Medicine*, 2nd ed.; Doyle, D., Hanke, G.W.C., MacDonald, N., Eds.; Oxford University Press: New York, NY, USA, 1998; pp. 1057–1075.

100. Heese, O.; Vogeler, E.; Martens, T.; Schnell, O.; Tonn, J.C.; Simon, M.; Schramm, J.; Krex, D.; Schackert, G.; Reithmeier, T.; et al. End-of-life caregivers' perception of medical and psychological support during the final weeks of glioma patients: A questionnaire-based survey. *Neuro-Oncology* **2013**, *15*, 1251–1256. [CrossRef] [PubMed]

© 2018 by the authors. Licensee MDPI, Basel, Switzerland. This article is an open access article distributed under the terms and conditions of the Creative Commons Attribution (CC BY) license (http://creativecommons.org/licenses/by/4.0/).

bioengineering

MDPI

Review

Preclinical Models of Pediatric Brain Tumors—Forging Ahead

Tara H.W. Dobson [1,*] and Vidya Gopalakrishnan [1,2,3,4,5]

[1]   Department of Pediatrics, University of Texas, M.D. Anderson Cancer Center, Houston, TX 77030, USA
[2]   Department of Molecular & Cellular Oncology, University of Texas, M.D. Anderson Cancer Center,
      Houston, TX 77030, USA; vgopalak@mdanderson.org
[3]   Brain Tumor Center, University of Texas, M.D. Anderson Cancer Center, Houston, TX 77030, USA
[4]   Center for Cancer Epigenetics, University of Texas, M.D. Anderson Cancer Center, Houston, TX 77030, USA
[5]   Graduate School of Biomedical Sciences UT-Health Science Center, Houston, TX 77030, USA
*   Correspondence: thdobson@mdanderson.org; Tel. +1-713-745-8291

Received: 21 August 2018; Accepted: 27 September 2018; Published: 2 October 2018

Abstract: Approximately five out of 100,000 children from 0 to 19 years old are diagnosed with a brain tumor. These children are treated with medication designed for adults that are highly toxic to a developing brain. Those that survive are at high risk for a lifetime of limited physical, psychological, and cognitive abilities. Despite much effort, not one drug exists that was designed specifically for pediatric patients. Stagnant government funding and the lack of economic incentives for the pharmaceutical industry greatly limits preclinical research and the development of clinically applicable pediatric brain tumor models. As more data are collected, the recognition of disease sub-groups based on molecular heterogeneity increases the need for designing specific models suitable for predictive drug screening. To overcome these challenges, preclinical approaches will need continual enhancement. In this review, we examine the advantages and shortcomings of in vitro and in vivo preclinical pediatric brain tumor models and explore potential solutions based on past, present, and future strategies for improving their clinical relevancy.

Keywords: preclinical; in vitro models; animal models; pediatric brain tumors

---

1. Introduction

Preclinical models contributed to significant advancements in the field of oncology during the past century, particularly the mechanistic understanding of tumor initiation and progression, resulting in improved diagnostic and prognostic assessment of patients. However, studies to identify novel therapeutics has limited value in the clinic. Mastering the design of models that efficiently progress preclinical studies from basic science and target identification, to translational research and drug discovery, to Food and Drug Administration (FDA)-approved drugs with clinical benefit is an ongoing endeavor. Only one in 5000 compounds go on to receive FDA approval. The success rate of new cancer therapies in phase I clinical trials is a dismal 3.4%, the lowest among major diseases [1]. Novel cancer therapeutics that make it to the clinic only have a one in ten chance of helping patients, and even then, the maximum increase in survival is 5.8 months [2,3]. To date, not one of these drugs was developed specifically for pediatric patients, despite the undeniable understanding that pediatric tumors are not the same as adult cancers. Not only are the tumors diverse, but there is the additional factor of treating a growing child—for children with central nervous system (CNS) tumors, we are talking about a developing brain. For now, pediatric cancer researchers are limited to studies with repurposed compounds. Therefore, attention should be placed on designing the best models possible to accurately predict which therapeutics (1) can target heterogeneous human tumors, (2) have minimal toxic effects on a developing child, and, for brain cancers, (3) has the ability to cross the blood–brain barrier. We anticipate this will require comprehensive knowledge of past observations, benefits and limitations of current studies, and innovative ideas with the potential to move pediatric research forward.

2. History of Preclinical Animal Models of Pediatric Brain Cancer

Modeling cancers has been around since 1915 when the first mice were exposed to coal tar in an attempt to create skin tumors in a laboratory [4]. It was not until 1939 when Seligman and Shear successfully utilized a carcinogen, methylcholanthrene, to generate a model for studying brain cancer [5]. Methylcholanthrene pellets were intracranially implanted into 20 mice, resulting in 11 gliomas and two meningeal fibrosarcomas [5]. They used this model for growth studies and were able to subcutaneously passage the tumors [4]. By the 1960s, with a greater understanding of carcinogenesis, studies of synthetic chemicals as possible mutagens commenced [4]. The synthetic carcinogen, N-nitrosourea, as well as derivatives ethylnitrosourea (ENU) and N-nitrosomethylurea, was found to induce brain tumors in rats [4,6]. Initial studies determined younger animals were more susceptible to tumor development compared to older rats after ENU injections. Transplacental oncogenic properties were identified after pups born from rats injected with ENU during pregnancy developed tumors [6]. An additional study, seven years later, reported a single intravenous injection of ENU at gestational day 20 resulted in tumors in 100% of the 25 offspring [7]. An astonishing total of 102 neural tumors developed in these pups consisting of oligodendrogliomas, astrocytomas, mixed gliomas, anaplastic gliomas, ependymomas, one meningioma, and neurinomas [7]. In 1936, intracerebral inoculation of RNA Rous sarcoma virus (RSV)-1, an avian retrovirus, induced intracerebral sarcoma in adult chickens [8]. By 1964, Rabotti et al. demonstrated the importance of tumor location after discovering intracerebral inoculations of RSV injected into different areas of hamster brains resulted in different tumor types [9]. A study in beagles using RSV demonstrated 100% penetrance in newborn pups after intracerebral injections. The developed tumors were shown to respond to chemotherapies from the time, (BCNU (carmustine) and CCNU (lomustine), but not cyclophosphamide [10]. These results mimicked observations in the clinic with human brain tumor patients [4]. Together with other studies, they demonstrated variations in dosage, age, strain, and overall health of the animal effected tumor type, malignancy characteristics, and location and rate of incidence [11]. In addition, this research contributed to important discoveries including the identification and characterization of blood–brain barrier defects, as well as aberrant endothelial and perivascular spread [4,12,13].

The development of brain tumor models most commonly used today began over thirty years ago. At this time, the first models of pediatric brain tumors were generated. In 1983, Neely et al. attempted to transplant 85 different pediatric tumors into immunodeficient animals, but had varying degrees of success [14]. Establishing tumors from brain in nude mice went very poorly, as did tissue from lymphoid benign tumors [14]. Simultaneously, cell lines derived from pediatric brain tumors including medulloblastoma (MB), atypical teratoid rhabdoid tumor (AT/RT), and high-grade gliomas (HGG) were developed, and, by the late 1980s, transplantable models were successfully established [15–21]. As the ability to genetically manipulate the mouse genome improved, mice with altered oncogenes and tumor suppressor genes were engineered [22–25]. These early patient-derived xenografts (PDX) and genetically engineered mouse models (GEMM) gained promise for use in biomedical analysis as they resembled human tumors and responded to therapeutics similar to clinical observations [4,12,13].

Historically, the initial inspiration to generate animal models was to test the effects of treatment on the tumors, but these older versions quickly proved to be unreliable. Tumor type and location lacked consistency with the chemically induced models [4]. By the same token, using viruses to generate brain tumors resulted in inconsistent tumor growth characteristics due to uneven distribution of recombinant viruses [4,12,13]. However, this pioneering research was instrumental for the characterization of CNS tumorigenesis, and greatly contributed to our understanding of the importance of tumor location, the surrounding microenvironment, and cells of origin [4,12,13]. This basic knowledge inspired many molecular and genetic studies of neural tumors that, in turn, resulted in improved diagnostics such as risk stratification, epidemiologic studies, and therapeutic intervention in the clinic [12]. The shortcomings of these models were hardly failures, nor should they be viewed as such. They actually laid the groundwork for current guidelines used to define a good preclinical model such as high rate of incidence, short latency, molecular and histopathological properties of human disease, and the ability to predict human response to treatment [26].

3. Applications to the Clinic: The Good, the Bad, and the Ugly

Direct clinically relevant goals of preclinical studies are to determine preliminary efficacy, toxicity, and pharmacokinetic and safety information of novel drugs in order to support the development of human trials [26,27]. In vitro and in vivo methods can be used to determine if the anticipated therapeutic characteristics of a new drug required for clinical success, such as absorption, disposition, metabolism, elimination, and toxicity, are present [28]. Every approach is individually limited; thus, understanding the benefits and barriers is of utmost importance when designing a study. Utilization of a strategic combination is imperative for successfully translating any findings to the clinic.

Cultured studies were performed for decades to explore the underlying biological mechanisms of cancer development. In vitro studies were used to identify the genetic and epigenetic changes in cancer cells that contribute to tumor initiation [29–33]. In addition, these tools are very useful to predict the response and resistance of cancers to different treatments [28,34–36]. Pharmacological high-throughput drug screening is readily applied to identify and evaluate potential therapeutics [29]. Moreover, diagnostically, the use of high-throughput omic analysis showed that molecular signatures of pediatric CNS tumors better predict patient outcome compared to histopathology alone [37–41].

Model systems used for in vitro research include mouse or human-derived cell lines and primary cells such as tumor stem cells or neurosphere cultures [29,30,42–44]. Cell lines that maintain the genetic perturbations of the primary tumors from which they were derived are ideal models [29]. Approximately 60 cell lines were generated from pediatric brain tumors including ependymomas (EPN), MB, HGG, and AT/RT (extensively reviewed by Xu et al.) [33]. A majority of these lines represent MB, which consists of four genetically distinct subgroups designated Wingless (WNT), Sonic Hedgehog (SHH), Group 3, and Group 4 [41]. In addition to molecular characteristics, prognoses vary between these types of MBs making modeling for specific subgroups a necessity. A newly designed bioinformatic classification tool predicted MB cells DAOY, D425, ONS-76, D283, D341, PFSK-1, D384, D458 to be WNT or SHH subtypes [45]. However, unlike WNT and SHH tumors, many of the lines are *MYC*-amplified like Group 3 MB [41]. Only 12 MB cells line were shown through various analyses to be specifically affiliated with a subgroup [46]. Except for one Group 4, these are all SHH or Group 3 cell lines [46]. A potential explanation for this discrepancy is the requirement of serum for the propagation of many pediatric brain tumor cell lines [33,47,48]. The use of serum to culture primary tumor cells leads to terminal cell differentiation and a homogenous cell population that acquires multiple molecular aberrations, so that, over time, the resulting cell lines differ significantly from the corresponding primary tumors [43]. Even though genetic and phenotypic drift and a lack of heterogeneity occur with cell lines, they are still commonly used in both genetic and pharmacological studies [49]. Cell lines are easy to work with because they are robust, grow rapidly, are easily modified, and can be stored long-term [44].

Some of the challenges with cell lines were addressed after the discovery of tumor stem cells when neurosphere cultures from EPN, MB, and astrocytoma pediatric brain tumors were first generated [30,31]. Neurospheres are cultured in the absence of serum and maintain tumor heterogeneity, making them an attractive alternative to serum cultured cells [49,50]. Spheres are heterogeneous and are good candidates for proliferation and differentiation assays despite technical challenges [30,32,42,43,49]. Although studies using sphere lines are a truer reflection of primary tumor behavior compared to cell lines, sphere lines are finicky and rarely maintained by long-term culturing [32,42]. Propagating tumor cells under these conditions select for neural stem-cell-like cells.

Many pediatric brain tumors arise from more lineage committed cells; therefore, generating neurosphere cultures from these tumors is not optimal [51,52]. To date, a single Group 4 MB neurosphere cell line (CHLA-01-MED), and a subsequent line from a malignant pleural effusion from the same patient (CHLA-01R-MED) were established and are commercially available (American Type Culture Collection (ATCC): CRL–3021 and CRL-3034, respectively) [49,53]. However new serum-free models, such as the HGG neurosphere lines by Wenger et al., which can be cultured up to 30 passages, continue to be generated [54].

Cell and sphere lines are helpful tools when addressing biological and mechanistic questions. The cost, convenience, and ease of in vitro techniques make them an attractive choice during initial inquiries of novel drugs, particularly for high-throughput testing of therapeutic agents [29]. After positive preliminary results, clinically applicable research of pediatric brain tumors often requires additional evaluation of novel chemotherapeutics and adjunctive therapies in animal models.

In vivo studies are considered the gold standard in the pharmaceutical industry when addressing drug safety and efficacy questions to support human testing [55]. Many organisms are used in different fields, but the most common host for pediatric brain research is the mouse. Like in vitro, in vivo techniques possess both advantageous and limiting characteristics, all of which require consideration when designing a preclinical trial. Anatomical characteristics of a host animal to cogitate include extracellular matrix molecules, cytokines/growth factors, endothelium, tissue-specific progenitor cells, and immune cells [12]. Depositories like JAX Mice and Services (part of the International Mouse Strain Resource) carry commercially available strains with varying types of immunodeficiency, and some can tolerate clinically relevant levels of radiation. As previously mentioned, the two major variations of murine models used in pediatric brain studies are PDXs and GEMMs.

Xenograft modeling refers to the implantation of primary tumor cells/tissue, or tumor-derived cell or sphere lines, classically under the skin (subcutaneous) or, more recently, in the native tumor site (orthotopic) of a host animal (Figure 1). Models generated by implantation of either mouse or human-derived cell lines, are inexpensive and predictable with rapid tumor growth, yet lack tumor heterogeneity and are rarely infiltrative [12,44,49,56]. Syngeneic xenografts, where the host and transplantable materials are genetically and immunologically matched, will contain a realistic microenvironment and intact immune system [12,13,57]. On the other hand, human cell-line models, while lacking both benefits, better represent human biology. Unlike cell lines, transplanting primary sphere lines or tumor tissue allows for maintained heterogeneity [4,16,47,48,51,57–61]. There is also the possibility of a pertinent microenvironment in the low-passage models of primary tumor transplants as tumor stroma may be present [26,44]. For pediatric brain cancer, mainly orthotopic PDXs are applied to drug efficacy comparisons between primary and corresponding metastatic tumors [62]. After a clinical trial, they are also used to identify molecular changes between the pre- and post-treatment tumors of non-responders [12,34].

Of course, there are many caveats to contend with. Results may only be reflective of the individual tumors samples. Multiple passages select the most aggressive cells, depleting heterogeneity [12,44,49,56]. PDXs from patient tumors are limited by the amount of available material and variable engraftment rates (tumor tissue from a patient with poor prognosis is the most efficacious); furthermore, a sufficient cohort comes with high cost and high effort [12,44,49,56]. As with any xenograft, implantation can disrupt the cell–matrix interactions and the blood–brain barrier, and lack tumor initiation [12]. Fortunately, these last three limitations can be bypassed using GEMMs.

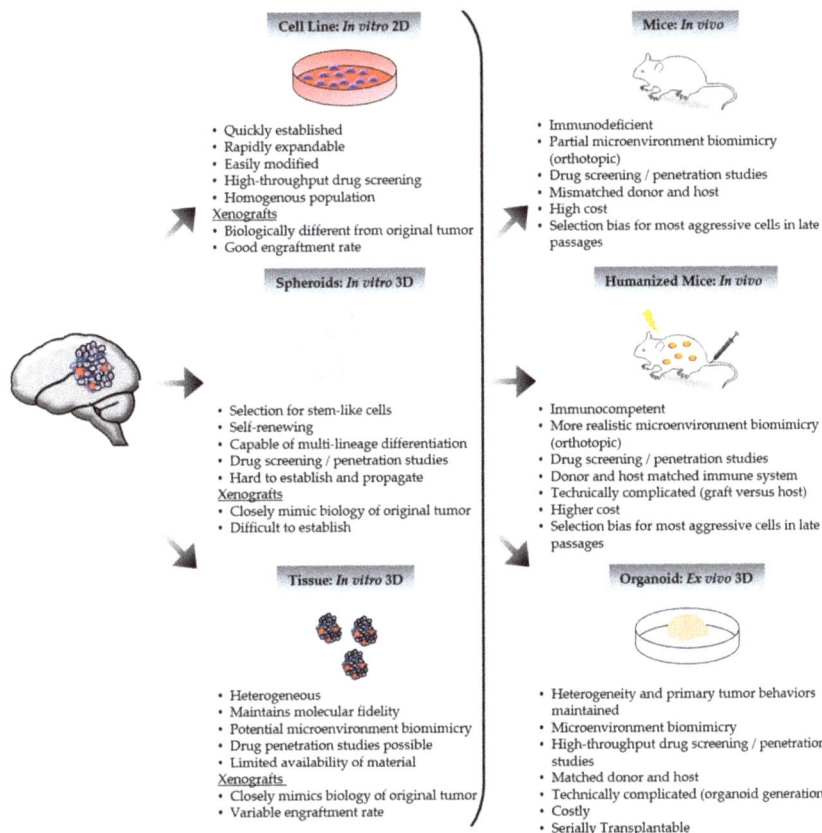

Figure 1. Characteristics of patient-derived model systems. Human tumor tissue can be processed to generate in vitro, ex vivo, and in vivo models. Xenografts can yield from tumor-derived in vitro models, as depicted above, or directly from patient tumor tissue. Represented tumor features include cancer cells (purple), cancer stem cells (blue), and blood vessels (red).

GEMMs are engineered by retroviral, proviral, or chemically induced mutations, including the addition of foreign DNA or transgenes, or by targeted mutations such as truncated or deleted gene knock-outs or amplified gene knock-ins [13,49]. Genes and mutations studied with GEMMs are often associated with a specific human disease and are highly controllable. They may be conventional or conditional to control spatial and temporal expression through the use of systems like Cre-*LoxP* or tetracycline-controlled transcriptional activation (Tet-off/on) [13,49]. GEMMs are often used to investigate individual genetic events in pathogenesis or tumor cell of origin [57]. They can also be designed to imitate rare subgroups. The majority of pediatric brain tumor GEMMs model MBs, mainly representing SHH MB tumors (extensively reviewed by Neumann et al.) [44]. After the link between Gorlin syndrome and aberrant SHH signaling in MB was discovered, many MB GEMMs were generated by modifying SHH signaling genes, such as *Ptch*, *Smo*, or *Sufu*, often in combination with deletion of *Trp53* or cyclin-dependent kinases [63–65]. Progenitors of the lower rhombic lip were identified as the cell of origin for WNT MBs after the development of the WNT MB GEMM which is genetically engineered to overexpress *Ctnnb1* and a *Pik3ca* mutation in combination with *Trp53* knock-out [66]. Compared to WNT and SHH, less is known about the drivers of Group 3 and Group 4 MBs, resulting in few models [44]. The GTML (*Glt1-tTA/TRE-MYCN-Luc*) model, a MB

GEMM, develops tumors that closely resemble Group 3, but WNT, SHH, and Group 4 can also arise [52,64,67]. There are currently no specific Group 4 MB GEMMs available [44]. Unlike xenografts, GEMMs recapitulate tumor initiation and progression in the presence of the native immune system, blood–brain barrier, cell–matrix interactions, and microenvironment, making them attractive models for targeted therapeutics and drug delivery studies [13]. However, like PDXs, the development and characterization of GEMMs are time-consuming endeavors with a high cost. Tumor penetrance can be incomplete with unpredictable growth [13,44]. GEMM tumors may lack heterogeneity, potentially limiting the ability to recapitulate the genetic complexity seen in human tumors [12,44,49]. Importantly, in the absence of a conserved, phenotypic response to genetic aberrations, the tumorigenesis and drug response seen in GEMMs often differ from humans [13].

Each preclinical pediatric brain tumor model has advantages and pitfalls. Despite many attempts, these models have yet to improve therapeutic options for pediatric patients. Not a single molecular targeting drug was added to a standard treatment protocol for pediatric brain tumors and only one, everolimus, has FDA approval for treatment of nonoperable subependymal giant cell astrocytoma [68]. However, many candidates were identified and are currently under clinical assessment. A promising example is the identification of numerous targets of the SHH pathway in SHH-driven MB. Inhibitors developed against smoothened (SMO) went through phase I and II trials [69]. An interesting finding in the phase II studies showed a wide range of drug efficacy in patients. These results paralleled the discovery of tumor diversity by omic evaluation, which was recapitulated in murine models [70]. Post-clinical research showed a subset of the non-responders had mutations downstream of SMO in the SHH pathway, suggesting a different inhibitor of downstream targets such as zinc finger protein (GLI) may benefit those patients [69]. Research targeting BRD4 (bromodomain-containing protein 4), a BET (bromodomain and extra terminal domain) protein that regulates *GLI* transcription, demonstrated both PDX and GEMM-models of Hedgehog-driven tumors (including SHH MBs and AT/RTs), with genetic perturbations resulting in resistance to SMO inhibitors, responded to treatment with the BRD4 inhibitor JQ1 [71]. This is an excellent example of "from bench to bedside and back again" and exemplifies the need to improve development of preclinical studies as the continued identification of tumor variations and subgroups will require the design of more specific models.

4. Promising Progress

A key expectation for a good preclinical model is the ability to predict human response to treatment. Although traditional models are extremely valuable tools, interpreting the results of model studies to predict clinical response may be the biggest challenge to overcome. A good start to addressing this problem is the reevaluation of preclinical strategies. Due to clear differences between mice and humans, ideas to humanize preclinical trials need to be in the forefront of future model designing.

To diminish time and cost, plus obvious ethical reasons, cultured studies need to be revisited. In vitro research is often mistakenly minimalized in terms of clinical value. However, pharmaceutical companies heavily rely on these methods, and even demonstrated that, at times, product assessment is more translatable with in vitro studies compared to in vivo [27,28,55,72]. In response, the development of some interesting and innovative ideas evolved in recent years. The explosion of multiple omic methodologies resulted in numerous readily available primary tumor datasets including those from the International Cancer Genome Consortium (ICGC), The Cancer Genome Atlas (TCGA), and, for pediatrics, Therapeutically Applicable Research to Generate Effective Treatments (TARGET). These data can be compared to omic studies of human cell lines in order to identify one that best represents a specific cancer subtype [29,73]. Resources are continuously updated and could be used to amplify the generation of new, diverse, and biologically relevant tumor cell lines with potential to enhance the success of drug discovery studies and create more applicable human-derived xenograft models [33,55,74,75].

Another exciting new avenue of in vitro modeling is three-dimensional (3D) growth cultures. Bingle et al. recently determined that 3D prototypes better represent neuroblastoma physiology compared to two-dimensional (2D) cultures [76]. With the use of bioreactor systems, they demonstrated how pertinent biological and physiological conditions, such as shear stress, compound flux, and removal of metabolites, could be studied with 3D cultures [76]. A third promising cell-derived tool for preclinical therapeutic research is the organoid. Organoids are used to study physiological processes in a setting closely resembling endogenous cell organization and organ architecture. An elegant new study by Ogawa et al. demonstrated that a cerebral organoid could recapitulate glioblastoma development via CRISPR/Cas9 (clustered regularly interspaced short palindromic repeats/CRISPR-associated protein 9) oncogene manipulation or transplantation of organoid- or adult-patient-derived glioblastoma cell lines [77]. These are great examples of what can be accomplished. For reasons mentioned above, in vitro studies should always be the first choice when results are equivalent to, or even better than, those obtained using animals. However, for now, most in vitro studies require additional testing before any clinical trial could be considered.

Building better animal models is a continually evolving pursuit as no animal can be a perfect human exemplar. A promising trend, first used to study blood cancers, but recently gaining popularity in solid tumor research, is the humanization of mice. Humanized mice are immunocompromised animals that receive human bone marrow to reconstitute the immune system. Compared to other models, these animals provide a more realistic tumor microenvironment and tumor heterogeneity with the potential for better drug response prediction. Humanized mouse models of diffuse intrinsic pontine glioma (DIPG), HGG, and high-risk MB were utilized to demonstrate that a DNA-damaging reagent, currently in trials to treat lung cancer, called 6-thio-2'deoxyguanosine could cross the blood–brain barrier and selectively enter tumor cells [78]. The downside to working with these animals, like with any murine model, is that they are expensive and technically complicated by the risk of xenogeneic graft-versus-host responses; however, they are a step closer to recapitulating human cancers in animals.

An ideal alternative to mice may be zebrafish (Table 1). The cancer genomes between zebrafish and humans are highly conserved [79]. Patient tumors can be transplanted to form PDXs. Genetically modified models can be generated by adding single or combined mutations [80]. Both approaches could potentially be used for compound toxicity screens, as well as time- and cost-efficient high-throughput in vivo analyses, although the pharmacokinetics of most drugs in zebrafish has yet to be determined [81]. Zebrafish are translucent, allowing direct imaging of tumor behavior with just a microscope [79]. Moreover, a single female can produce up to 200 embryos a day [79]. Successful transplants of mouse-derived ependymoma, glioma, and choroid plexus carcinoma were achieved [81]. More recently, an oligoneural/NB-FOXR2 (NB forkhead box R2) CNS primitive neuroectodermal tumor type-specific model was generated by activating N-RAS in Olig2+/ Sox10+ (oligodendrocyte transcription factor /sex-determining region Y box 10) oligoneural precursor cells of embryonic zebrafish [82]. Mesenchymal glioblastoma and low-grade glioma models were also generated [83]. Whether zebrafish tumors are better representations of human tumors than mice is still unknown, but early studies indicate they will be valuable tools for future research.

Table 1. Zebrafish in cancer modeling.

	Benefits			Limitations
Translational relevance: similarities between humans and zebrafish	Largely conserved development and signaling pathways	Function of innate and adaptive immune cells is highly conserved	Over 80% of human disease-related genes present	Whole-genome duplication (more than one ortholog for some human genes) may interfere with genetic studies
Xenograft models [79,81,84]	Can be generated from human, mouse, or zebrafish tumors	No rejection due to immature adaptive immune system in larvae	Recapitulates parental tumor behaviors including proliferation, survival, invasion, and dissemination	Molecular interactions between transplanted human or mouse tumors and zebrafish cells unclear

Table 1. Cont.

	Benefits			Limitations
Genetically engineered zebrafish models (GEZMs) [82,85,86]	Easy genetic manipulation—injection into one-cell-stage larvae possible	Fast development	Comparable histology to human cancers	Tumor initiation and progression studies hindered by lack of a functional adaptive immune system in early-stage models
Drug studies	Easy, cost-effective, and high scalability	Ease of imaging and high-throughput screening with transparent larvae	High degree of conservation of metabolic enzymes between human and zebrafish larvae	Pharmacokinetic processes still unclear

5. Concluding Remarks

In order to improve the clinical relevancy of preclinical research, we must acknowledge the major hurdles impeding our progress. Preclinical versus clinic studies and endpoints do not resemble each other. Slowing of tumor progression as a measurement of disease response to treatment is defined in the clinic as complete response, partial response, and overall increase in survival versus in the lab as tumor growth inhibition and tumor growth delay. The latter criteria do not directly correlate to an overall increase in survival. The discrepancy between these definitions of success and failure impede both discovery and extensive testing of prospective therapeutics. The challenge to bridge basic science and clinical communication is still in front of us. Multidisciplinary collaborations are being developed worldwide as a potential solution, and time will tell if they are enough to maximize preclinical–clinical synergy. Then, there are the funding limitations. As discussed in this review, no single model can currently replicate the development, diversity, or drug responsiveness of a human tumor; furthermore, a dismal percentage of compounds identified by preclinical cancer research become beneficial therapeutics in the clinic. Therefore, a question deserving serious consideration is whether investing more upfront for preclinical research would lead to better model design and improve drug discovery. Regardless of the solution, increased efficiency in the lab is necessary to ultimately decrease the current number of unsuccessful and expensive human trials.

Author Contributions: T.D. and V.G. designed the concept of the review; T.D. performed the literature research, wrote the manuscript, and designed the figures.

Funding: This work was supported by grants from the National Institutes of Health (5R01-NS-079715-01 and 5R03NS077021-01), American Cancer Society (RSG-09-273-01-DDC), Cancer Prevention Research Institute of Texas (CPRIT-RP150301), and Rally Foundation for Childhood Cancers to V.G., and the UT MD Anderson Cancer Center-CCE Scholar Program to T.D.

Acknowledgments: The authors wish to thank Amanda R. Haltom, Jyothishmathi Swaminathan, and Javiera Bravo-Alegria for their critical reading of the manuscript and insightful comments and suggestions.

Conflicts of Interest: The authors declare no conflicts of interest.

References

1. Wong, C.H.; Siah, K.W.; Lo, A.W. Estimation of clinical trial success rates and related parameters. *Biostatistics* **2018**, *00*, 1–14. [CrossRef] [PubMed]
2. Strand, A.D.; Girard, E.; Olson, J.M. *Patient Derived Tumor Xenograft Models: Promise, Potential and Practice*; Uthamanthil, R., Tinkey, P., Eds.; Elsvier Inc.: London, UK, 2017.
3. Davis, C.; Naci, H.; Gurpinar, E.; Poplavska, E.; Pinto, A.; Aggarwal, A. Availability of evidence of benefits on overall survival and quality of life of cancer drugs approved by European Medicines Agency: Retrospective cohort study of drug approvals 2009–13. *BMJ* **2017**, *359*, 0959–8138. [CrossRef] [PubMed]
4. Peterson, D.L.; Sheridan, P.J.; Brown, W.E., Jr. Animal models for brain tumors: Historical perspectives and future directions. *J. Neurosurg.* **1994**, *80*, 865–876. [CrossRef] [PubMed]
5. Seligman, A.M.; Shear, M.J.; Alexander, L. Studies in carcinogenesis: VIII. Experimental production of brain tumors in mice with methylcholanthrene. *Am. J. Cancer* **1939**, *37*, 364–395.
6. Druckrey, H.; Ivankovic, S.; Preussmann, R. Teratogenic and carcinogenic effects in the offspring after single injection of ethylnitrosourea to pregnant rats. *Nature* **1966**, *210*, 1378–1379. [CrossRef] [PubMed]

7. Koestner, A.; Swenberg, J.A.; Wechsler, W. Transplacental production with ethylnitrosourea of neoplasms of the nervous system in Sprague-Dawley rats. *Am. J. Pathol.* **1971**, *63*, 37–56. [PubMed]
8. Vazquez-Lopez, E. On the growth of Rous sarcoma inoculated into the brain. *Am. J. Cancer* **1939**, *29*, 29–55. [CrossRef]
9. Rabotti, G.F.; Raine, W.A. Brain tumours induced in hamsters inoculated intracerebrally at birth with rous sarcoma virus. *Nature* **1964**, *204*, 898–899. [CrossRef] [PubMed]
10. Cuatico, W.; Cho, J.R.; Spiegelman, S. Molecular evidence for a viral etiology of human CNS tumors. *Acta Neurochir.* **1976**, *35*, 149–160. [CrossRef] [PubMed]
11. Yoshida, J.; Cravioto, H. Nitrosourea-induced brain tumors: An in vivo and in vitro tumor model system. *J. Natl. Cancer Inst.* **1978**, *61*, 365–374. [PubMed]
12. Huszthy, P.C.; Daphu, I.; Niclou, S.P.; Stieber, D.; Nigro, J.M.; Sakariassen, P.O.; Miletic, H.; Thorsen, F.; Bjerkvig, R. In vivo models of primary brain tumors: Pitfalls and perspectives. *Neuro. Oncol.* **2012**, *14*, 979–993. [CrossRef] [PubMed]
13. Simeonova, I.; Huillard, E. In vivo models of brain tumors: Roles of genetically engineered mouse models in understanding tumor biology and use in preclinical studies. *Cell. Mol. Life Sci.* **2014**, *71*, 4007–4026. [CrossRef] [PubMed]
14. Neely, J.E.; Ballard, E.T.; Britt, A.L.; Workman, L. Characteristics of 85 pediatric tumors heterotransplanted into nude mice. *Exp. Cell. Biol.* **1983**, *51*, 217–227. [CrossRef] [PubMed]
15. Friedman, H.S.; Burger, P.C.; Bigner, S.H.; Trojanowski, J.Q.; Wikstrand, C.J.; Halperin, E.C.; Bigner, D.D. Establishment and characterization of the human medulloblastoma cell line and transplantable xenograft D283 Med. *J. Neuropathol. Exp. Neurol.* **1985**, *44*, 592–605. [CrossRef] [PubMed]
16. Jacobsen, P.F.; Jenkyn, D.J.; Papadimitriou, J.M. Establishment of a human medulloblastoma cell line and its heterotransplantation into nude mice. *J. Neuropathol. Exp. Neurol.* **1985**, *44*, 472–485. [CrossRef] [PubMed]
17. Keles, G.E.; Berger, M.S.; Srinivasan, J.; Kolstoe, D.D.; Bobola, M.S.; Silber, J.R. Establishment and characterization of four human medulloblastoma-derived cell lines. *Oncol. Res.* **1995**, *7*, 493–503. [PubMed]
18. Yachnis, A.T.; Neubauer, D.; Muir, D. Characterization of a primary central nervous system atypical teratoid/rhabdoid tumor and derivative cell line: Immunophenotype and neoplastic properties. *J. Neuropathol. Exp. Neurol.* **1998**, *57*, 961–971. [CrossRef] [PubMed]
19. Friedman, H.S.; Burger, P.C.; Bigner, S.H.; Trojanowski, J.Q.; Brodeur, G.M.; He, X.M.; Wikstrand, C.J.; Kurtzberg, J.; Berens, M.E.; Halperin, E.C.; et al. Phenotypic and genotypic analysis of a human medulloblastoma cell line and transplantable xenograft (D341 Med) demonstrating amplification of c-myc. *Am. J. Pathol.* **1988**, *130*, 472–484. [PubMed]
20. Wasson, J.C.; Saylors, R.L.; Zeltzer, P.; Friedman, H.S.; Bigner, S.H.; Burger, P.C.; Bigner, D.D.; Look, A.T.; Douglass, E.C.; Brodeur, G.M. Oncogene amplification in pediatric brain tumors. *Cancer Res.* **1990**, *50*, 2987–2990. [PubMed]
21. Pietsch, T.; Scharmann, T.; Fonatsch, C.; Schmidt, D.; Ockler, R.; Freihoff, D.S.; Albrecht, O.D.; Wiestler, P.; Zeltzer, H. Characterization of five new cell lines derived from human primitive neuroectodermal tumors of the central nervous system. *Cancer Res.* **1994**, *54*, 3278–3287. [PubMed]
22. Goodrich, L.V.; Milenkovic, L.; Higgins, K.M.; Scott, M.P. Altered neural cell fates and medulloblastoma in mouse patched mutants. *Science* **1997**, *277*, 1109–1113. [CrossRef] [PubMed]
23. Lee, Y.; McKinnon, P.J. DNA ligase IV suppresses medulloblastoma formation. *Cancer Res.* **2002**, *62*, 6395–6399. [PubMed]
24. Marino, S.; Vooijs, M.H.; Der Gulden, V.; Jonkers, J.; Berns, A. Induction of medulloblastomas in p53-null mutant mice by somatic inactivation of Rb in the external granular layer cells of the cerebellum. *Gene. Dev.* **2000**, *14*, 994–1004. [PubMed]
25. Wetmore, C.; Eberhart, D.E.; Curran, T. The normal patched allele is expressed in medulloblastomas from mice with heterozygous germ-line mutation of patched. *Cancer Res.* **2000**, *60*, 2239–2246. [PubMed]
26. Perrin, S. Preclinical research: Make mouse studies work. *Nature* **2014**, *507*, 423–425. [CrossRef] [PubMed]
27. Zhang, D.; Luo, G.; Ding, X.; Lu, C. Preclinical experimental models of drug metabolism and disposition in drug discovery and development. *Acta Pharm. Sin. B* **2012**, *2*, 549–561. [CrossRef]
28. Li, A.P. Preclinical in vitro screening assays for drug-like properties. *Drug Discov. Today Technol.* **2005**, *2*, 179–185. [CrossRef] [PubMed]
29. Goodspeed, A.; Heiser, L.M.; Gray, J.W.; Costello, J.C. Tumor-derived cell lines as molecular models of cancer pharmacogenomics. *Mol. Cancer Res.* **2016**, *14*, 3–13. [CrossRef] [PubMed]

30. Hemmati, H.D.; Nakano, I.; Lazareff, J.A.; Masterman-Smith, M.; Geschwind, D.H.; Bronner-Fraser, M.; Kornblum, H.I. Cancerous stem cells can arise from pediatric brain tumors. *Proc. Natl. Acad. Sci. USA* **2003**, *100*, 15178–15183. [CrossRef] [PubMed]

31. Singh, S.K.; Clarke, I.D.; Terasaki, M.; Bonn, V.E.; Hawkins, C.; Squire, J.; Dirks, P.B. Identification of a cancer stem cell in human brain tumors. *Cancer Res.* **2003**, *63*, 5821–5828. [PubMed]

32. Suslov, O.N.; Kukekov, V.G.; Ignatova, T.N.; Steindler, D.A. Neural stem cell heterogeneity demonstrated by molecular phenotyping of clonal neurospheres. *Proc. Natl. Acad. Sci. USA* **2002**, *99*, 14506–14511. [CrossRef] [PubMed]

33. Xu, J.; Erdreich-Epstein, A.; Gonzalez-Gomez, I.; Melendez, E.Y.; Smbatyan, G.; Moats, R.A.; Rosol, M.; Biegel, J.A.; Reynolds, C.P. Novel cell lines established from pediatric brain tumors. *J. Neurooncol.* **2012**, *107*, 269–280. [CrossRef] [PubMed]

34. Houghton, P.J.; Morton, C.L.; Tucker, C.D.; Payne, E.; Favours, C.; Cole, R.; Gorlick, E.A.; Kolb, W.; Zhang, R.; Lock, H.; et al. The pediatric preclinical testing program: Description of models and early testing results. *Pediatr. Blood Cancer* **2007**, *49*, 928–940. [CrossRef] [PubMed]

35. Morfouace, M.; Nimmervoll, B.; Boulos, N.; Patel, Y.T.; Shelat, A.; Freeman, B.B.; Robinson, G.W.; Wright, K.; Gajjar, A.; Stewart, C.F.; et al. Preclinical studies of 5-fluoro-2'-deoxycytidine and tetrahydrouridine in pediatric brain tumors. *J. Neurooncol.* **2016**, *126*, 225–234. [CrossRef] [PubMed]

36. Sewing, A.C.P.T.; Lagerweij, D.G.; van Vuurden, M.H.; Meel, S.J.E.; Veringa, A.M.; Carcaboso, P.J.; Gaillard, W.; Peter Vandertop, P.; Wesseling, D.; Noske, G.J.L.; et al. Preclinical evaluation of convection-enhanced delivery of liposomal doxorubicin to treat pediatric diffuse intrinsic pontine glioma and thalamic high-grade glioma. *J. Neuros-Pediatr.* **2017**, *19*, 518–530. [CrossRef] [PubMed]

37. Bae, J.M.; Won, J.K.; Park, S.H. Recent advancement of the molecular diagnosis in pediatric brain tumor. *J. Korean Neurosurg. Soc.* **2018**, *61*, 376–385. [CrossRef] [PubMed]

38. Gajjar, A.; Pfister, S.M.; Taylor, M.D.; Gilbertson, R.J. Molecular insights into pediatric brain tumors have the potential to transform therapy. *Clin. Cancer Res.* **2014**, *20*, 5630–5640. [CrossRef] [PubMed]

39. Kool, M.; Korshunov, A.; Remke, M.; Jones, D.T.; Schlanstein, M.; Northcott, P.A.; Cho, Y.J.; Koster, J.; Schouten-van Meeteren, A.; van Vuurden, D.; et al. Molecular subgroups of medulloblastoma: An international meta-analysis of transcriptome, genetic aberrations, and clinical data of WNT, SHH, Group 3, and Group 4 medulloblastomas. *Acta Neuropathol.* **2012**, *123*, 473–484. [CrossRef] [PubMed]

40. Pomeroy, S.L.; Tamayo, P.; Gaasenbeek, M.; Sturla, L.M.; Angelo, M.; McLaughlin, M.E.; Kim, J.Y.; Goumnerova, L.C.; Black, P.M.; Lau, C.; et al. Prediction of central nervous system embryonal tumour outcome based on gene expression. *Nature* **2002**, *415*, 436–442. [CrossRef] [PubMed]

41. Taylor, M.D.; Northcott, P.A.; Korshunov, A.; Remke, M.; Cho, Y.J.; Clifford, S.C.; Eberhart, C.G.; Parsons, D.W.; Rutkowski, S.; Gajjar, A.; et al. Molecular subgroups of medulloblastoma: The current consensus. *Acta Neuropathol.* **2012**, *123*, 465–472. [CrossRef] [PubMed]

42. Bez, A.; Corsini, E.; Curti, D.; Biggiogera, M.; Colombo, A.; Nicosia, R.F.; Pagano, S.F.; Parati, E.A. Neurosphere and neurosphere-forming cells: Morphological and ultrastructural characterization. *Brain Res.* **2003**, *993*, 18–29. [CrossRef] [PubMed]

43. Lee, J.; Kotliarova, S.; Kotliarov, Y.; Li, A.; Su, Q.; Donin, N.M.; Pastorino, S.; Purow, B.W.; Christopher, N.; Zhang, W.; Park, J.K.; Fine, H.A. Tumor stem cells derived from glioblastomas cultured in bFGF and EGF more closely mirror the phenotype and genotype of primary tumors than do serum-cultured cell lines. *Cancer Cell.* **2006**, *9*, 391–403. [CrossRef] [PubMed]

44. Neumann, J.E.; Swartling, F.J.; Schuller, U. Medulloblastoma: Experimental models and reality. *Acta Neuropathol.* **2017**, *134*, 679–689. [CrossRef] [PubMed]

45. Gendoo, D.M.; Smirnov, P.; Lupien, M.; Haibe-Kains, B. Personalized diagnosis of medulloblastoma subtypes across patients and model systems. *Genomics* **2015**, *106*, 96–106. [CrossRef] [PubMed]

46. Ivanov, D.P.; Coyle, B.; Walker, D.A.; Grabowska, A.M. In vitro models of medulloblastoma: Choosing the right tool for the job. *J. Biotechnol.* **2016**, *236*, 10–25. [CrossRef] [PubMed]

47. Dietl, S.; Schwinn, S.; Dietl, S.; Riedel, S.; Deinlein, F.; Rutkowski, S.; von Bueren, A.O.; Krauss, J.; Monoranu, T.; Schweitzer, G.H.; et al. Wolfl, MB3W1 is an orthotopic xenograft model for anaplastic medulloblastoma displaying cancer stem cell- and Group 3-properties. *BMC Cancer* **2016**, *16*, 115. [CrossRef] [PubMed]

48. Milde, T.; Lodrini, M.; Savelyeva, L.; Korshunov, A.; Kool, M.; Brueckner, L.M.; Antunes, A.S.; Oehme, I.; Pekrun, A.; Pfister, S.M.; et al. HD-MB03 is a novel Group 3 medulloblastoma model demonstrating sensitivity to histone deacetylase inhibitor treatment. *J. Neurooncol.* **2012**, *110*, 335–348. [CrossRef] [PubMed]

49. Sandén, E. Experimental models of pediatric brain tumors. Establishment, immunophenotyping and clinical implications. Ph.D. Thesis, Lund University, Scania, Sweden, January 2016.

50. Zhou, Z.; Luther, N.; Singh, R.; Boockvar, J.A.; Souweidane, M.M.; Greenfield, J.P. Glioblastoma spheroids produce infiltrative gliomas in the rat brainstem. *Childs Nerv. Syst.* **2017**, *33*, 437–446. [CrossRef] [PubMed]

51. Monje, M.; Mitra, S.S.; Freret, M.E.; Raveh, T.B.; Kim, J.; Masek, M.; Attema, J.L.; Li, G.; Haddix, T.; Edwards, M.S.; et al. Hedgehog-responsive candidate cell of origin for diffuse intrinsic pontine glioma. *Proc. Natl. Acad Sci. USA* **2011**, *108*, 4453–4458. [CrossRef] [PubMed]

52. Swartling, F.J.; Savov, V.; Persson, A.I.; Chen, J.; Hackett, C.S.; Northcott, P.A.; Grimmer, M.R.; Lau, J.; Chesler, L.; Perry, A.; et al. Distinct neural stem cell populations give rise to disparate brain tumors in response to N-MYC. *Cancer Cell.* **2012**, *21*, 601–613. [CrossRef] [PubMed]

53. Xu, J.; Margol, A.S.; Shukla, A.; Ren, X.; Finlay, J.L.; Krieger, M.D.; Gilles, F.H.; Couch, F.J.; Aziz, M.; Fung, E.T.; et al. Disseminated Medulloblastoma in a Child with Germline BRCA2 6174delT Mutation and without Fanconi Anemia. *Front. Oncol.* **2015**, *5*, 191. [CrossRef] [PubMed]

54. Wenger, A.; Larsson, S.; Danielsson, A.; Elbaek, K.J.; Kettunen, P.; Tisell, M.; Sabel, M.; Lannering, B.; Nordborg, C.; Schepke, E.; et al. Stem cell cultures derived from pediatric brain tumors accurately model the originating tumors. *Oncotarget* **2017**, *8*, 18626–18639. [CrossRef] [PubMed]

55. Polli, J.E. In vitro studies are sometimes better than conventional human pharmacokinetic in vivo studies in assessing bioequivalence of immediate-release solid oral dosage forms. *AAPS J.* **2008**, *10*, 289–299. [CrossRef] [PubMed]

56. Mak, I.W.; Evaniew, N.; Ghert, M. Lost in translation: Animal models and clinical trials in cancer treatment. *Am. J. Transl. Res.* **2014**, *6*, 114–118. [PubMed]

57. Sreedharan, S.; Maturi, N.P.; Xie, Y.; Sundstrom, A.; Jarvius, M.; Libard, S.; Alafuzoff, I.; Weishaupt, H.; Fryknas, M.; Larsson, R.; et al. Uhrbom, Mouse Models of Pediatric Supratentorial High-grade Glioma Reveal How Cell-of-Origin Influences Tumor Development and Phenotype. *Cancer Res.* **2017**, *77*, 802–812. [CrossRef] [PubMed]

58. Baxter, P.A.; Lin, Q.; Mao, H.; Kogiso, M.; Zhao, X.; Liu, Z.; Huang, Y.; Voicu, H.; Gurusiddappa, S.; Su, J.M.; et al. Silencing BMI1 eliminates tumor formation of pediatric glioma CD133+ cells not by affecting known targets but by down-regulating a novel set of core genes. *Acta Neuropathol. Com.* **2014**, *2*, 160. [CrossRef] [PubMed]

59. Girard, E.; Ditzler, S.; Lee, D.; Richards, A.; Yagle, K.; Park, J.; Eslamy, H.; Bobilev, D.; Vrignaud, P.; Olson, J. Efficacy of cabazitaxel in mouse models of pediatric brain tumors. *Neuro. Oncol.* **2015**, *17*, 107–115. [CrossRef] [PubMed]

60. Hennika, T.; Hu, G.; Olaciregui, N.G.; Barton, K.L.; Ehteda, A.; Chitranjan, A.; Chang, C.; Gifford, A.J.; Tsoli, M.; Ziegler, D.S.; et al. Pre-Clinical Study of Panobinostat in Xenograft and Genetically Engineered Murine Diffuse Intrinsic Pontine Glioma Models. *PLoS ONE* **2017**, *12*, e0169485. [CrossRef] [PubMed]

61. Zhao, X.; Zhao, Y.J.; Lin, Q.; Yu, L.; Liu, Z.; Lindsay, H.; Kogiso, M.; Rao, P.; Li, X.N.; Lu, X. Cytogenetic landscape of paired neurospheres and traditional monolayer cultures in pediatric malignant brain tumors. *Neuro Oncol.* **2015**, *17*, 965–977. [CrossRef] [PubMed]

62. Stacchiotti, S.; Saponara, M.; Frapolli, R.; Tortoreto, M.; Cominetti, D.; Provenzano, S.; Negri, T.; Dagrada, G.P.; Gronchi, A.; Colombo, C.; et al. Patient-derived solitary fibrous tumour xenografts predict high sensitivity to doxorubicin/dacarbazine combination confirmed in the clinic and highlight the potential effectiveness of trabectedin or eribulin against this tumour. *Eur. J. Cancer* **2017**, *76*, 84–92. [CrossRef] [PubMed]

63. Ayrault, O.; Zindy, F.; Rehg, J.; Sherr, C.J.; Roussel, M.F. Two tumor suppressors, p27Kip1 and patched-1, collaborate to prevent medulloblastoma. *Mol. Cancer Res.* **2009**, *7*, 33–40. [CrossRef] [PubMed]

64. Poschl, J.; Stark, S.; Neumann, P.; Grobner, S.; Kawauchi, D.; Jones, D.T.; Northcott, P.A.; Lichter, P.; Pfister, S.M.; Kool, M.; et al. Genomic and transcriptomic analyses match medulloblastoma mouse models to their human counterparts. *Acta Neuropathol.* **2014**, *128*, 123–136. [CrossRef] [PubMed]

65. Uziel, T.; Zindy, F.; Xie, S.; Lee, Y.; Forget, A.; Magdaleno, S.; Rehg, J.E.; Calabrese, C.; Solecki, D.; Eberhart, C.G.; et al. The tumor suppressors Ink4c and p53 collaborate independently with Patched to suppress medulloblastoma formation. *Genes Dev.* **2005**, *19*, 2656–2667. [CrossRef] [PubMed]

66. Gibson, P.; Tong, Y.; Robinson, G.; Thompson, M.C.; Currle, D.S.; Eden, C.; Kranenburg, T.A.; Hogg, T.; Poppleton, H.; Martin, J.; et al. Subtypes of medulloblastoma have distinct developmental origins. *Nature* **2010**, *468*, 1095–1099. [CrossRef] [PubMed]

67. Hill, R.M.; Kuijper, S.; Lindsey, J.C.; Petrie, K.; Schwalbe, E.C.; Barker, K.; Boult, J.K.; Williamson, D.; Ahmad, Z.; Hallsworth, A.; et al. Combined MYC and P53 defects emerge at medulloblastoma relapse and define rapidly progressive, therapeutically targetable disease. *Cancer Cell.* **2015**, *27*, 72–84. [CrossRef] [PubMed]

68. Turner, S.G.; Peters, K.B.; Vredenburgh, J.J.; Desjardins, A.; Friedman, H.S.; Reardon, D.A. Everolimus tablets for patients with subependymal giant cell astrocytoma. *Expert Opin. Pharmacother.* **2011**, *12*, 2265–2269. [CrossRef] [PubMed]

69. Rimkus, T.K.; Carpenter, R.L.; Qasem, S.; Chan, M.; Lo, H.W. Targeting the Sonic Hedgehog Signaling Pathway: Review of Smoothened and GLI Inhibitors. *Cancers* **2016**, *8*, 22. [CrossRef] [PubMed]

70. Kool, M.; Jones, D.T.; Jager, N.; Northcott, P.A.; Pugh, T.J.; Hovestadt, V.; Piro, R.M.; Esparza, L.A.; Markant, S.L.; Remke, M.; et al. Genome sequencing of SHH medulloblastoma predicts genotype-related response to smoothened inhibition. *Cancer Cell.* **2014**, *25*, 393–405. [CrossRef] [PubMed]

71. Tang, Y.; Gholamin, S.; Schubert, S.; Willardson, M.I.; Lee, A.; Bandopadhayay, P.; Bergthold, G.; Masoud, S.; Nguyen, B.; Vue, N.; et al. Epigenetic targeting of Hedgehog pathway transcriptional output through BET bromodomain inhibition. *Nat. Med.* **2014**, *20*, 732–740. [CrossRef] [PubMed]

72. Johnson, J.I.; Decker, S.; Zaharevitz, D.; Rubinstein, L.V.; Venditti, J.M.; Schepartz, S.; Kalyandrug, S.; Christian, M.; Arbuck, S.; Hollingshead, M.; et al. Relationships between drug activity in NCI preclinical in vitro and in vivo models and early clinical trials. *Br. J. Cancer* **2001**, *84*, 1424–1431. [CrossRef] [PubMed]

73. Kraljevic, S.; Stambrook, P.J.; Pavelic, K. Accelerating drug discovery. *EMBO Rep.* **2004**, *5*, 837–842. [CrossRef] [PubMed]

74. Samson, K. New Campaign Seeks to Stimulate Research on Pediatric Brain Cancers. *Oncol. Times* **2016**, *38*, 14–15. [CrossRef]

75. Zhao, X.; Liu, Z.; Yu, L.; Zhang, Y.; Baxter, P.; Voicu, H.; Gurusiddappa, S.; Luan, J.; Su, J.M.; Leung, H.C.; et al. Global gene expression profiling confirms the molecular fidelity of primary tumor-based orthotopic xenograft mouse models of medulloblastoma. *Neuro Oncol.* **2012**, *14*, 574–583. [CrossRef] [PubMed]

76. Bingel, C.; Koeneke, E.; Ridinger, J.; Bittmann, A.; Sill, M.; Peterziel, H.; Wrobel, J.K.; Rettig, I.; Milde, T.; Fernekorn, U.; et al. Three-dimensional tumor cell growth stimulates autophagic flux and recapitulates chemotherapy resistance. *Cell. Death Dis.* **2017**, *8*, e3013. [CrossRef] [PubMed]

77. Ogawa, J.; Pao, G.M.; Shokhirev, M.N.; Verma, I.M. Glioblastoma Model Using Human Cerebral Organoids. *Cell. Rep.* **2018**, *23*, 1220–1229. [CrossRef] [PubMed]

78. Sengupta, S.; Sobo, M.; Lee, K.; Senthil, K.S.; White, A.R.; Mender, I.; Fuller, C.; Chow, L.M.L.; Fouladi, M.; Shay, J.W.; et al. Induced Telomere Damage to Treat Telomerase Expressing Therapy-Resistant Pediatric Brain Tumors. *Mol. Cancer Ther.* **2018**, *17*, 1504–1514. [CrossRef] [PubMed]

79. Casey, M.J.; Modzelewska, K.; Anderson, D.; Goodman, J.; Boer, E.F.; Jimenez, L.; Grossman, D.; Stewart, R.A. Transplantation of Zebrafish Pediatric Brain Tumors into Immune-competent Hosts for Long-term Study of Tumor Cell Behavior and Drug Response. *J. Vis. Exp.* **2017**, *123*, e55712. [CrossRef] [PubMed]

80. Kirchberger, S.; Sturtzel, C.; Pascoal, S.; Distel, M. Quo natas, Danio?-Recent Progress in Modeling Cancer in Zebrafish. *Front. Oncol.* **2017**, *7*, 186. [CrossRef] [PubMed]

81. Eden, C.J.; Ju, B.; Murugesan, M.; Phoenix, T.N.; Nimmervoll, B.; Tong, Y.; Ellison, D.W.; Finkelstein, D.; Wright, K.; Boulos, N.; et al. Orthotopic models of pediatric brain tumors in zebrafish. *Oncogene* **2015**, *34*, 1736–1742. [CrossRef] [PubMed]

82. Modzelewska, K.; Boer, E.F.; Mosbruger, T.L.; Picard, D.; Anderson, D.; Miles, R.R.; Kroll, M.; Oslund, W.; Pysher, T.J.; Schiffman, J.D.; Jensen, R.; et al. MEK Inhibitors Reverse Growth of Embryonal Brain Tumors Derived from Oligoneural Precursor Cells. *Cell. Rep.* **2016**, *17*, 1255–1264. [CrossRef] [PubMed]

83. Mohanty, S.; Chen, Z.; Li, K.; Morais, G.R.; Yerneni, J.K.; Pisani, L.; Chin, F.T.; Mitra, S.; Cheshier, S.; Chang, E.; et al. A novel theranostic strategy for MMP-14-expressing glioblastomas impacts survival. *Mol. Cancer Ther.* **2017**, *16*, 1909–1921. [CrossRef] [PubMed]

84. Welker, A.M.; Jaros, B.D.; Puduvalli, V.K.; Imitola, J.; Kaur, B.; Beattie, C.E. Standardized orthotopic xenografts in zebrafish reveal glioma cell-line-specific characteristics and tumor cell heterogeneity. *Dis. Model. Mech.* **2016**, *9*, 199–210. [CrossRef] [PubMed]

85. Ju, B.; Chen, W.; Orr, B.A.; Spitsbergen, J.M.; Jia, S.; Eden, C.J.; Henson, H.E.; Taylor, M.R. Oncogenic KRAS promotes malignant brain tumors in zebrafish. *Mol. Cancer* **2015**, *14*, 18. [CrossRef] [PubMed]

86. Mayrhofer, M.; Gourain, V.; Reischl, M.; Affaticati, P.; Jenett, A.; Joly, J.S.; Benelli, M.; Demichelis, F.; Poliani, P.L.; Sieger, D.; et al. A novel brain tumour model in zebrafish reveals the role of YAP activation in MAPK- and PI3K-induced malignant growth. *Dis. Model. Mech.* **2017**, *10*, 15–28. [CrossRef] [PubMed]

© 2018 by the authors. Licensee MDPI, Basel, Switzerland. This article is an open access article distributed under the terms and conditions of the Creative Commons Attribution (CC BY) license (http://creativecommons.org/licenses/by/4.0/).

bioengineering

MDPI

Review

Embryonal Tumors of the Central Nervous System in Children: The Era of Targeted Therapeutics

David E. Kram [1,*,†], Jacob J. Henderson [2,†], Muhammad Baig [3], Diya Chakraborty [3],
Morgan A. Gardner [4], Subhasree Biswas [3,5] and Soumen Khatua [3]

1   Section of Pediatric Hematology-Oncology, Department of Pediatrics, Wake Forest School of Medicine,
    Winston Salem, NC 27157, USA
2   Pape Family Pediatric Research Institute, Department of Pediatrics, Oregon Health and Science University,
    Portland, OR 97239, USA; hendejac@ohsu.edu
3   Department of Pediatrics, MD Anderson Cancer Center, Houston, TX 77030, USA;
    MBaig@mdanderson.org (M.B.); DChakraborty@mdanderson.org (D.C.); subhabis@gmail.com (S.B.);
    SKhatua@mdanderson.org (S.K.)
4   Department of Pediatrics, Wake Forest School of Medicine, Winston Salem, NC 27157, USA;
    magardne@wakehealth.edu
5   Humanitas University Medical School, 20090 Milan, Italy
*   Correspondence: dkram@wakehealth.edu; Tel.: +1-336-716-4085; Fax: +1-336-716-3010
†   These authors contributed equally the writing and preparation of the article.

Received: 17 August 2018; Accepted: 12 September 2018; Published: 23 September 2018

Abstract: Embryonal tumors (ET) of the central nervous system (CNS) in children encompass a wide clinical spectrum of aggressive malignancies. Until recently, the overlapping morphological features of these lesions posed a diagnostic challenge and undermined discovery of optimal treatment strategies. However, with the advances in genomic technology and the outpouring of biological data over the last decade, clear insights into the molecular heterogeneity of these tumors are now well delineated. The major subtypes of ETs of the CNS in children include medulloblastoma, atypical teratoid rhabdoid tumor (ATRT), and embryonal tumors with multilayered rosettes (ETMR), which are now biologically and clinically characterized as different entities. These important developments have paved the way for treatments guided by risk stratification as well as novel targeted therapies in efforts to improve survival and reduce treatment burden.

Keywords: central nervous system; embryonal tumors; children; medulloblastoma; ATRT; ETMR; molecular biology; targeted therapeutics

1. Introduction

Over the past decade, a surge genomic and epigenomic data on embryonal tumors of the central nervous system (CNS) in children has dramatically advanced the understanding of tumor biology, paving the way toward improved diagnostic, reclassification, and therapeutic approaches to these formidable malignancies [1]. These new molecular phenotypes have been incorporated in the 2016 World Health Organization (WHO) classification of CNS tumors, creating a major shift in paradigm of the classification of embryonal tumors of the CNS. The 2016 WHO classification integrated genetic information to already-existing histopathological data and has enabled more precise classification and diagnosis of these tumors [2]. This is most evident in the recognition of the molecular subtyping of medulloblastoma (MB), each subtype carrying unique demographics and clinical outcomes. Until recently, all non-medulloblastoma embryonal tumors were encompassed under the umbrella of CNS-primitive neuroectodermal tumors (PNET). Revelations from several molecular-profiling and methylation assay studies now show that a range of distinctly biologically heterogeneous tumors exist that are now known as non-medulloblastoma embryonal tumors of the CNS. The main non-medulloblastoma embryonal tumors include atypical teratoid/rhabdoid tumors (ATRT) and embryonal tumors with multilayered rosettes (ETMR). Other CNS embryonal tumors include the following morphological subtypes: medulloepithelioma, CNS neuroblastoma, CNS ganglioneuroblastoma, and ependymoblastoma [3]. All variants of embryonal tumors are now well-defined entities with varying molecular biology and prognosis. The critical issue is proper diagnosis of these entities, which would allow tailored therapy, including intensifying treatment for aggressive variants and de-escalating therapy for those tumors with better prognoses, in order to achieve long-term cures and minimizing treatment-related toxicity. The following section will discuss the major subtypes of pediatric embryonal tumors, which include MB, ATRT, and ETMR, their molecular biology, and the insights it provides in developing targeted therapies.

2. Medulloblastoma

2.1. Introduction

MB is the most common pediatric CNS malignancy, accounting for approximately 20% of all pediatric CNS tumors [4,5]. Despite the conventional aggressive multimodal therapeutic approach involving surgery, radiotherapy, and chemotherapy, approximately one-third of patients with MB die from their disease [6]. Survivors often experience long-term sequelae, including neurocognitive deficits, endocrinopathies, and secondary malignancies [7–10]. Integrative genomic and methylomic studies over the past 15 years have shown that MB is not a single entity, but rather a heterogeneous group of diseases with unique clinical, molecular, and prognostic characteristics [11–14]. The internationally accepted subgroups of MB have been termed wingless (WNT), sonic hedgehog (SHH), Group 3, and Group 4, which have been adopted in the 2016 WHO classification of central nervous system tumors [2,15]. These efforts to characterize MB have revolutionized our understanding of its pathogenesis and response to treatment, and will hopefully lead to improvements in survival and survivors' quality of life.

2.2. Molecular Subgroups

MB subgroups exhibit distinct biological characteristics, and recent application of advanced algorithms for integrative genomic analysis has highlighted heterogeneity even within these subgroups, which may help inform patient stratification in future trials (Figure 1) [16].

WNT: WNT MBs are characterized by activation of the WNT signaling pathway, commonly caused by somatic mutations in the CTNNB1 gene, results in stabilization of the beta-catenin protein [17,18]. WNT MBs are the rarest subgroup, almost never present with metastatic disease, and carry the most favorable prognosis of all the subgroups [12,15,19].

SHH: SHH MBs are characterized by hyperactivation of the SHH signaling pathway, often due to germline or somatic mutations or amplifications in components of the SHH pathway. SHH MBs represent approximately 30% of all MBs and are associated with a markedly variable prognosis [20–25]. Of note, upwards of a quarter of patients with SHH tumors have an underlying germline mutation including TP53 or BRCA, which ought to prompt clinicians to consider genetic counseling consultation in all patients diagnosed with SHH MB [26].

Group 3: A unifying pathway has not been identified in Group 3 MBs, as these tumors contain few recurrent somatic and germline mutations aside from amplification of the proto-oncogene MYC and isochromosome 17q [14,22,23,27]. Group 3 MBs account for about 25% of all MBs, affect almost exclusively infants and young children, and have the worst prognosis.

Group 4: The underlying biological abnormalities characteristic of this subgroup are the least understood. Isochromosome 17q is commonly found, but unlike in Group 3 MB, this feature is not associated with poor prognosis [28]. Similar to SHH and Group 3 MBs, there is intertumoral heterogeneity in Group 4 MB [16]. Group 4 is the most common subtype and, despite their common presentation with metastatic disease, overall prognosis of this subgroup is intermediate.

	WNT	SHH	Group 3	Group 4
Percentage	10%	30%	25%	35%
Age				
Sex ratio (M:F)	1:1	1:1	2:1	2.5:1
Location	CP/CPA angle	Cerebellar hemispheres	Midline - enhancement	Midline + enhancement
Metastasis at diagnosis	5-10%	15-20%	40-50%	35-40%
Genetic alteration	CTNNB1, DDX3X, SMARCA4, TP53	PTCH1, SMO, SUFU, TP53, GLI2, MYCN	GLI1, GFL1B, MYC, OTX2, SMARCA4	KDM6A, SNCAIP, CDK6, MYCN
Cytogenetic aberrations	Monosomy 6	3q gain, 9q, 10q, 17p loss	i17q, 1q, 7, 18 gain, 10q, 11, 17p loss	i17q, 7q, 18q gain, 8p, 11p, X loss
Potential targeted agents	De-escalation of therapy, axin inhibitors, PARPi	SMOi, GLI1/2i, PI3Ki, Aurora kinase and PLKi	BETi, CDKi, HDACi, PI3Ki, MTORi, WEE1i	CDKi, MYCNi, HDACi

Figure 1. Molecular subgroups of medulloblastoma with unique clinical, genetic, and prognostic characteristics. CP/CPA, cerebellar-pontine/cerebellar-pontine angle; SHH, sonic hedgehog; WNT, wingless and int.

2.3. Standard Therapy

Risk stratification: Historically, MB was classified as "average risk" or "high risk" based on age, presence of metastasis at diagnosis, and extent of residual tumor after resection. However, with the recognition of MB subgroups and their respective and independent impact on prognosis, a new risk-stratification schema has been proposed (Figure 2) [29].

	WNT	SHH	Group 3	Group 4
Low Risk (>90% Survival)	-All*			-Ch11 loss and non-metastatic
Standard Risk (75-90% Survival)		-Non-metastatic -TP53 wildtype -MYCN non-amp	-Non-metastatic and non-MYC-amp	-Non-metastatic and no Ch11 loss
High Risk (50-75% Survival)		-Metastatic (TP53 wildtype) -MYCN amp		-Metastatic
Very High Risk (<50% Survival)		-TP53 mutated	-Metastatic -MYC-amp**	

Figure 2. Current risk-stratification schema. * Metastatic WNT tumors are rare and have an unknown natural history. ** Nonmetastatic MYC-amplified tumors are likely very high risk and often recur in a metastatic pattern, but their upfront prognosis is not clear. Ch11, chromosome 11.

Standard treatments: Current therapy for MB consists of maximal tumor resection, craniospinal irradiation (age permitting), and multiagent chemotherapy. The historical goal was a gross total resection, but when controlling for molecular subgroup, the extent of resection may be less important [29]. Radiation is an effective component of MB treatment, though it comes with acute and long-term complications [10,30]. Lower doses may be utilized for certain patients, and conformal, intensity-modulated approaches, as well as electron- and proton-based therapies, can mitigate some toxicities [31–35]. Chemotherapy typically includes the cyclophosphamide/vincristine/cisplatin combination, while radiation-sparing approaches for infants employ combinations of high-dose chemotherapies, often with autologous stem-cell rescue [31,36].

2.4. Molecularly Targeted Therapy

WNT: WNT/β-catenin pathway overexpression is the hallmark of WNT MB, which may lend these tumors to be susceptible to axin or PARP inhibitors. Beyond targeted approaches, the favorable outcomes in patients with WNT MB have encouraged efforts aimed at de-escalation of first-line therapy. The Children's Oncology Group (COG) study ACNS1422, St. Jude's (SJ) study SJMB12, and the International Society of Paediatric Oncology's PNET 5 study all reduce the doses of craniospinal irradiation, while the Johns Hopkins J1403 trial was designed to eliminate it altogether [37–39].

SHH: Efforts to target the SHH pathway have focused on inhibiting the transmembrane receptor smoothened (SMO). Several SMO inhibitors have been introduced into clinical studies, including sonidegib and vismodegib, which have shown safety and improved progression-free survival when used as monotherapy, but both almost universally succumb to development of resistance [40–45]. Vismodegib is now being studied in combination with conventional therapy in an ongoing phase 2 study (SJMB12) [37]. Another compelling target is the downstream transcription factor, glioma-associated oncogene (GLI) [46].

Groups 3 and 4: The significant heterogeneity of the biological drivers of these MB subtypes has impaired the development of a clinically applicable targeted approach. However, the understanding that MYC-driven MBs rely on the kinase WEE1 to maintain cell-cycling viability has drawn interest, and an ongoing phase 2 trial (COG ADVL1312) is studying a WEE1 inhibitor in combination with irinotecan [47].

Nonspecific Subgroup Molecular Targeting and immunotherapy: Several other drivers of MB have been identified, though they do not yet fall into any single MB subgroup. The Notch signaling pathway has been implicated in MB tumorigenesis and may be targetable with γ-secretase inhibitors [48]. The PI3K/AKT/mTOR pathway is instrumental in multiple metabolic and survival pathways, and disturbances to this pathway have been implicated in MB [49,50]. Currently, the Pediatric MATCH trial is testing a dual PI3K/mTOR inhibitor in MBs with PI3K/TSC/mTOR mutations [51]. PI3K, histone deacetylase, and BET-bromodomain inhibitors, especially in MYC-amplified MB, show early preclinical promise and are being studied in an early-phase study (PBTC-026) [52–54]. Cyclin-dependent kinase 6 (CDK6) is commonly mutated in SHH, Group 3, and Group 4 MB [55]. Inhibitors of CDK4/6 are being studied in ongoing early-phase trials (PBTC-042 and SJDAWN) [56]. Activation of the RAS/MEK/ERK pathway has been observed in MBs, and targeting this pathway is being studied utilizing EGFR, VEGF-A, and MEK inhibitors [57–61].

Like many other pediatric tumors, the potential to harness the immune system's innate antitumor effect to treat MB is being increasingly appreciated. Early studies have shown that most MB tumors have only nominal immune-cell infiltration and tumor cells have absent or deficient antigen-presenting machinery [62,63]. Despite the apparent lack of tumor surveillance, there is growing evidence that checkpoint inhibition, immunization, or viral therapy may be of benefit [64–66].

2.5. Conclusion

Over the past 15 years, our understanding of MB has greatly increased, enabling the classification of subgroups. Subgroup-based risk stratification offers the potential to improve overall survival while reducing treatment burden in lower risk patients. The continuous discovery of biological diversity within MB subgroups, however, makes risk assignment and selection of appropriate patients for novel therapies more challenging [20]. Furthermore, the genomic and transcriptomic profiling of a tumor is likely spatially variable and is dynamic with treatment. By profiling MBs as bulk samples, there is risk of masking potential intratumoral heterogeneity, missing data of MB stem-cell programs, and concealing information about nonmalignant cells of the tumor microenvironment. MB is a heterogeneous disease, and treating all patients with standard therapy is no longer acceptable. However, while the era of targeted therapies based on genetic perturbations has great promise, future clinical trials will require novel approaches to best pair potent therapies to susceptible tumors.

3. Atypical Teratoid Rhabdoid Tumors (ATRT)

3.1. Introduction

ATRTs represent a variant of embryonal tumors of the CNS affecting younger, predominantly male children. They account for 1–2% of CNS tumors in children, with a peak incidence of children aged less than 3 years. Two-thirds of these neoplasms occur in the cerebellum, commonly at the cerebellar-pontine angle (CPA), with a potential of multifocal or disseminated disease in 20% of patients at the time of diagnosis [67]. Prognosis remains poor, though recent revelations of biological data and improved understanding of the signaling pathways now offer some therapeutic optimism [68]. Recent treatment strategies have focused on using radiation and chemotherapy along with targeted therapeutics, which could improve overall prognosis [69,70].

3.2. Clinical Features and Diagnosis

Clinical presentation of ATRT largely depends on the location of the tumor and metastatic status. These tumors do not have extraordinary neuroimaging characteristics; their appearances are quite similar to medulloblastoma. Their appearance is often heterogeneous with high cellularity, containing a distinct band of wavelike enhancement; they may have intratumoral hemorrhage and peripherally located cysts; they may also show patterns of restricted diffusion [71]. Recent findings suggest that MRI features may vary across different molecular subgroups of ATRT [72,73]. ATRT morphology is diverse, ranging from epithelial to mesenchymal to neuroepithelial, and sometimes containing all three. Classic rhabdoid cells are less prevalent in these tumors, and instead more commonly exhibit small-blue-round-cell tumor morphology, similar to medulloblastoma. Prior to the discovery of the SMARCB1 mutation as a specific molecular marker for ATRT, they were diagnosed under the same umbrella as PNET or medulloblastoma [74]. Currently accepted diagnostic criteria include biallelic loss and/or negative immunohistochemistry staining of SMARCB1 or SMARCA4 and their respective gene products hSNF/INI1/BAF47 and BRG1 [75,76]. Importantly, 20–35% of patients with ATRT are found to carry biallelic germline alterations of SMARCB1 or SMARCA4. These patients have rhabdoid tumor predisposition syndrome, with a propensity to develop intra- and extracranial aggressive rhabdoid tumors at young age [77,78]. Thus, all patients with ATRT should undergo genetic counseling and testing for the presence of a germline mutation.

3.3. Molecular Era of ATRT

The sole recurrent genetic alterations in these tumors are biallelic mutations of SMARCB1 (INI1, SNF5, BAF47) or, rarely, SMARCA4, both of which are members of the SWItch/Sucrose Nonfermentable chromatin-remodeling complex [79]. Approximately 20% of the 22q11.2 deletions include SMARCB1, of which 25% of the patients have partial deletion or duplication. The remaining chromosomal aberrations include single-base point mutations, frame shifts, or insertions [80].

High-resolution molecular studies have uncovered marked clinical and molecular heterogeneity in the relatively bland genome of ATRT. Varying molecular subgroups in ATRT were first identified by Birks et al., who showed high expression of bone morphologic protein (BMP) in a subgroup associated with shorter survival [81]. A subsequent larger study, using integrated analysis of clinical and transcriptional data, demonstrated two major subtypes with different clinical outcomes: the supratentorial tumors, which were characterized by neuronal differentiation and ASCL1 protein expression, correlated with improved survival outcome, while infratentorial tumors enriched with BMP signatures carried worse prognosis [82]. A recent study utilizing genetic, epigenetic, and transcriptional characterization subdivided ATRT into three methylation subgroups with varying demographics and molecular profiles. These are ATRT-SHH, ATRT-TYR, ATRT-MYC [79]. Figure 3 shows the molecular subtypes including demographics, SMARCB1 profile, epigenetic features, and therapeutic targets of interest. The generation of the first transgenic mouse model harboring temporal deletion/inactivation of SMARCB1 facilitated further understanding of the oncogenic events leading to ATRT formation. This study revealed that epigenetic mechanisms associated with hSNF5 loss drives these tumors and provided insights into the different targeted cells of origin that likely contribute to the heterogeneous nature of ATRT [83].

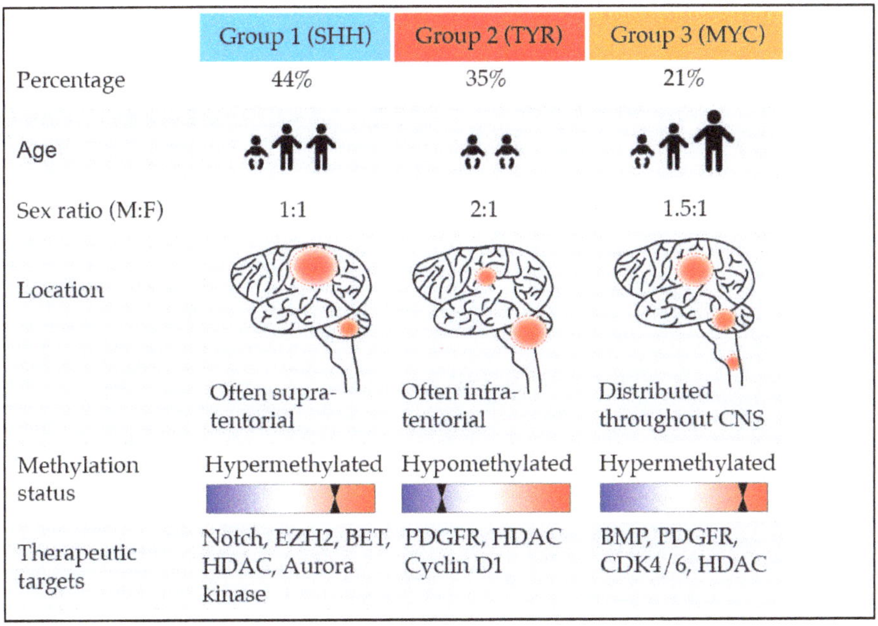

	Group 1 (SHH)	Group 2 (TYR)	Group 3 (MYC)
Percentage	44%	35%	21%
Age			
Sex ratio (M:F)	1:1	2:1	1.5:1
Location	Often supra-tentorial	Often infra-tentorial	Distributed throughout CNS
Methylation status	Hypermethylated	Hypomethylated	Hypermethylated
Therapeutic targets	Notch, EZH2, BET, HDAC, Aurora kinase	PDGFR, HDAC Cyclin D1	BMP, PDGFR, CDK4/6, HDAC

Figure 3. Molecular subgroups of atypical teratoid rhabdoid tumors (ATRT) and their clinical and epigenetic features, as well as therapeutic targets.

3.4. Current Treatment

Given the rarity of the disease and the diversity of treatment regimens historically employed to treat ATRT, no standard therapeutic approach exists. There seems to be improved survival for those patients with de novo ATRT treated with Intergroup Rhabdomyosarcoma Study (IRS) protocols and high-dose alkylating agents [84]. A phase II study by the Dana Farber Cancer Institute treated 20 patients with ATRT using a modified IRS III approach. Patients on this study had good outcomes, with an OS of 70% over two years, which was improved in comparison to other historical data [85]. A retrospective study by St. Jude Cancer Research Hospital included 31 children older than three years with ATRT who received craniospinal irradiation, high-dose alkylator-based chemotherapy, and showed that they fared better, with a two-year OS of $89 \pm 11\%$ [86]. The role of high-dose chemotherapy (HDC) followed by stem-cell rescue in studies by the Head Start and the Canadian

Brain Tumor Consortia showed improved outcomes with intensive chemotherapy regimens, though toxicity was not inconsequential [87,88]. ATRT registry data suggest that the extent of surgical resection correlates uniformly with improved survival [89]. The role of radiotherapy still remains unclear and in general has been deferred or dose-reduced in younger children with ATRT. However, recent studies have shown a trend towards increased survival time with the addition of radiotherapy, leading to its incorporation in various clinical trials [90,91]. The current COG ACNS0333 trial (NCI: NCT00653068) incorporates a combination of chemotherapy and radiation therapy, along with autologous stem-cell rescue in treating young patients with ATRT.

3.5. Newer Therapeutic Insights

It is now evident that ATRT is an epigenetically driven disease, and thus targeting oncogenic drivers in epigenetic machinery is now the key focus of developmental therapeutics against these malignancies. The finding that ATRT may exhibit increased expression of EZH2, and subsequent hypermethylation has led to preclinical testing of EZH2 inhibitors EPZ-6438, 3-deazaneplanocin, and tazemetostat, alone or in combination with other chemotherapies; the data are encouraging, showing some tumor responsiveness to this approach [92,93]. A phase I trial using tazemetostat monotherapy against hypertrimethylated tumors, including one with an INI1-deleted malignant rhabdoid tumor, showed promising results: the one ATRT-like patient achieved a partial response [94]. Upregulation of the Bromo/BET domain may also contribute to ATRT oncogenesis, and Bromo/BET inhibitors have shown some promise in preclinical models [95,96]. Histone deacetylase inhibitors, which are also epigenetic modifiers, have been successfully used to inhibit ATRT growth and increased sensitivity to radiation [97]. Aurora kinase A (AURKA) is highly expressed in ATRT because of the loss of the INI1 tumor-suppressor gene. Alisertib, an AURKA inhibitor, has been explored in ATRT with good outcomes in patients with progressive disease [69]. The association between INI1 loss and increased expression of cyclin D1 in nearly 80% of ATRTs led to testing cyclin D1 as a druggable target [98,99]. CDK4/6 inhibitor palbociclib, when used in combination with radiation in preclinical models, showed delayed growth of ATRT cells [100]. Following this finding, a clinical trial employed another CDK4/6 inhibitor, ribociclib, alone or in combination with conventional chemotherapy in a few patients with ATRT, and it has shown some initial success [101]. The highly heterogeneous nature of these tumors led to the possibility of exploring multitargeted tyrosine kinase inhibitors (TKI). A recent study showed second-generation TKIs nilotinib and dasatinib reducing cellular proliferation of these tumors, through downregulation of PDGFRβ, and more so in the clinically worse subtype 2 [79].

3.6. Conclusions

ATRT is an aggressive malignancy with poor overall survival in metastatic disease and in younger children. Though multiple therapeutic approaches have been pursued over the last two decades, prognosis has remained grim until recently. It is now suspected that epigenetic alterations may be important druggable targets, and a variety of trials testing this hypothesis are ongoing. Recent identification and description of molecular subclasses of ATRT should guide future risk-stratified and targeted therapeutic approaches with the intent of reducing toxicity and improving overall survival of these formidable tumors.

4. Embryonal Tumor with Multilayered Rosettes (ETMR)

4.1. Introduction

ETMR was first identified in 2009 as an aggressive embryonal tumor with a unique molecular phenotype occurring in younger children. Histologically, it represents a PNET with ependymoblastic rosettes and neuropil-like areas containing neurocytes and ganglion cells [102]. ETMR can exhibit three distinct histological patterns: embryonal tumors with abundant neuropil and true rosettes (ETANTR), ependymoblastoma (EBL), and medulloepithelioma (MEPL). The uniform amplification of C19MC in these three distinct histologies has led to the grouping of ETANTR, EBL, and MEPL into a single diagnostic entity known as ETMR [103,104].

4.2. Clinical Features and Diagnosis

Due to the rarity of ETMR and limited number of publications, robust demographics have not been defined. These tumors have a very aggressive clinical course, with a median survival of 12 months after diagnosis. ETMR usually occurs in children younger than four years, and is more common in girls, unlike the other CNS embryonal tumors, in which boys are equally or more commonly affected. Clinical features are determined by the location and extent of the tumor. Most are supratentorial in location, a few are infratentorial, and they are very rarely encountered in the spinal cord. Increased intracranial pressure, seizures, hemiparesis, cerebellar signs, cranial nerve palsies, and other neurologic deficits have been reported. In general, neuroimaging shows these tumors as large, demarcated, solid masses featuring patchy- or no-contrast enhancement with surrounding edema, often with significant mass effect. A minority of the reported cases have shown cystic components and microcalcifications [105,106]. Nearly 80% are localized at diagnosis, but even these often rapidly progress. The diagnosis of ETMR remains challenging with only histopathological reviews. LIN28A evaluation by immunohistochemistry remains confirmatory. Comparative genomic hybridization array for amplification of the microRNA cluster C19MC at locus 19q13.42 may further secure the diagnosis [107].

4.3. Current Treatment Strategies

Current treatment protocols include maximal-safe surgery with subsequent chemotherapy, often including HDC with stem-cell rescue and focal or craniospinal irradiation depending on the age of the patient and extent of the tumor. Conventional chemotherapeutic agents used include cyclophosphamide, methotrexate, vincristine, etoposide, and carboplatin. Due to the small number of ETMR patients reported in the literature, and variable treatment strategies rendered, it is difficult to draw robust conclusions on the benefits of different treatment regimens on clinical outcome. Though initial responses were seen in some cases, the aggressive nature of these tumors resulted in early progression or relapses and poor overall survival. Despite intensive multimodality therapy with aggressive surgery, chemotherapy, and radiation, five-year overall survival has been less than 10% [108]. Extent of resection seems to be a favorable prognostic factor, though, in another series, radiation therapy correlated with improved outcome [109]. The impact of HDC on long-term survival is still not clear due to the limited number of patients treated with this modality. Lack of biographical input of these tumors has undermined development of targeted therapies and improved survival. However, recent availability and studies on cell lines and preclinical models are now uncovering promising targets for these lethal tumors.

4.4. Molecular Characteristics and Therapeutic Insights

The uniformity of C19MC amplification in these tumors led to the restructuring of ETMR classification. Transcriptional signatures of C19MC-altered tumors reveal enrichment for early neural and pluripotency genes, including LIN28/LIN28B. This suggests the primitive nature of these malignancies and may explain their aggressive phenotype [110]. Global molecular studies identified LIN28A immunoexpression as a highly specific marker for ETMR, with a hallmark genomic amplification of the C19MC oncogenic miRNA cluster. Nearly 25% of ETMR tumors with the molecular features of C19MC alterations are devoid of classical rosette structures, indicative of intratumoral heterogeneity. Global methylation assay and exome-sequencing studies have failed to identify other recurrent alterations, reaffirming C19MC as the main oncogenic driver of these formidable tumors. Though highly suggestive of ETMR, enrichment of LIN28/LIN28B is not specific for ETMR, as this is also seen in ATRT and high-grade gliomas [111].

Lack of optimal preclinical ETMR models has historically precluded defining conventional or novel targeted therapies. However, recent establishment of ETMR cell lines has led to in vitro and in vivo preclinical drug screening. The MTOR pathway appears to be markedly overexpressed in these tumors and is sensitive to inhibition by mTOR inhibitors demonstrated in cell lines [104]. RNA sequencing identified recurrent gene fusions of C19MC miRNAs and neural-specific DNMT3B isoforms in these ETMR tumors. These findings suggest DNMT3B as an important downstream effector of the C19MC oncogene and a potential therapeutic target for these formidable tumors [110].

Recent development of patient-derived xenograft models is now also providing avenues to explore drug activity in vivo. A mouse model of ETMR shows parallel activation of SHH and WNT signaling. This confirms that coactivation of these pathways in mouse neural precursors is sufficient to induce ETMR tumors [112]. These neoplasms resemble human counterparts, both histopathologically and by global gene-expression analysis, suggesting this as a feasible mouse model for further biological interrogation. These tumors responded to the SHH inhibitor, arsenic trioxide, proposing downstream inhibition of SHH signaling as a therapeutic option for patients with ETMR. The role of combination chemotherapy in the ETMR xenograft mice model has been recently evaluated. A combination of topotecan or doxorubicin, along with methotrexate and vincristine, demonstrated longer survival of animals in the topotecan arm. This study also highlighted a potential role for epigenetic modifying drugs, such as panobinostat, and targeted drugs, such as the pololike kinase 1 inhibitor, volasertib, in these tumors [113]. These biological revelations provide the initial clues to profile treatment using a targeted therapeutic approach with conventional chemotherapeutic agents.

4.5. Conclusions

C19MC-altered tumors, now identified as ETMR, carry a distinct aggressive clinical course in the setting of conventional therapeutic strategies. However, improved understanding of biological/molecular data aided by preclinical studies now provides avenues to render novel targeted therapies. Efforts to identify genomic and epigenetic machinery driving C19MC-associated tumorigenesis should continue, which would help in developing more precise, less toxic, and optimal treatment for these potentially lethal tumors.

Author Contributions: D.E.K. and J.J.H. contributed equally to the preparation of the manuscript as first authors. All the authors contributed to the writing, editing, and literature review, including final edits of the manuscript. S.K. contributed to the overall supervision of the manuscript preparation.

Funding: No funding was received for the preparation of the manuscript.

Conflicts of Interest: The authors declare no conflict of interest.

References

1. Pomeroy, S.L.; Tamayo, P.; Gaasenbeek, M.; Sturla, L.M.; Angelo, M.; McLaughlin, M.E.; Kim, J.Y.; Goumnerova, L.C.; Black, P.M.; Lau, C.; et al. Prediction of central nervous system embryonal tumour outcome based on gene expression. *Nature* **2002**, *415*, 436–442. [CrossRef] [PubMed]
2. Louis, D.N.; Perry, A.; Reifenberger, G.; von Deimling, A.; Figarella-Branger, D.; Cavenee, W.K.; Ohgaki, H.; Wiestler, O.D.; Kleihues, P.; Ellison, D.W. The 2016 world health organization classification of tumors of the central nervous system: A summary. *Acta Neuropathol.* **2016**, *131*, 803–820. [CrossRef] [PubMed]
3. Sin-Chan, P.; Li, B.K.; Ho, B.; Fonseca, A.; Huang, A. Molecular classification and management of rare pediatric embryonal brain tumors. *Curr. Oncol. Rep.* **2018**, *20*, 69. [CrossRef] [PubMed]
4. McNeil, D.E.; Cote, T.R.; Clegg, L.; Rorke, L.B. Incidence and trends in pediatric malignancies medulloblastoma/primitive neuroectodermal tumor: A seer update. Surveillance epidemiology and end results. *Med. Pediatr. Oncol.* **2002**, *39*, 190–194. [CrossRef] [PubMed]
5. Weil, A.G.; Wang, A.C.; Westwick, H.J.; Ibrahim, G.M.; Ariani, R.T.; Crevier, L.; Perreault, S.; Davidson, T.; Tseng, C.H.; Fallah, A. Survival in pediatric medulloblastoma: A population-based observational study to improve prognostication. *J. Neuro-oncol.* **2017**, *132*, 99–107. [CrossRef] [PubMed]

6. Polkinghorn, W.R.; Tarbell, N.J. Medulloblastoma: Tumorigenesis, current clinical paradigm, and efforts to improve risk stratification. *Nature Clin. Pract. Oncol.* **2007**, *4*, 295–304. [CrossRef] [PubMed]
7. Holland, A.A.; Colaluca, B.; Bailey, L.; Stavinoha, P.L. Impact of attention on social functioning in pediatric medulloblastoma survivors. *Pediatr. Hematol. Oncol.* **2018**, *35*, 76–89. [CrossRef] [PubMed]
8. Kieffer, V.; Chevignard, M.P.; Dellatolas, G.; Puget, S.; Dhermain, F.; Grill, J.; Valteau-Couanet, D.; Dufour, C. Intellectual, educational, and situation-based social outcome in adult survivors of childhood medulloblastoma. *Dev. Neurorehabil.* **2018**, 1–8. [CrossRef] [PubMed]
9. Uday, S.; Murray, R.D.; Picton, S.; Chumas, P.; Raju, M.; Chandwani, M.; Alvi, S. Endocrine sequelae beyond 10 years in survivors of medulloblastoma. *Clin. Endocrinol.* **2015**, *83*, 663–670. [CrossRef] [PubMed]
10. Bavle, A.; Tewari, S.; Sisson, A.; Chintagumpala, M.; Anderson, M.; Paulino, A.C. Meta-analysis of the incidence and patterns of second neoplasms after photon craniospinal irradiation in children with medulloblastoma. *Pediatr. Blood Cancer* **2018**, *65*, e27095. [CrossRef] [PubMed]
11. Kool, M.; Koster, J.; Bunt, J.; Hasselt, N.E.; Lakeman, A.; van Sluis, P.; Troost, D.; Meeteren, N.S.; Caron, H.N.; Cloos, J.; et al. Integrated genomics identifies five medulloblastoma subtypes with distinct genetic profiles, pathway signatures and clinicopathological features. *PLoS ONE* **2008**, *3*, e3088. [CrossRef] [PubMed]
12. Cho, Y.J.; Tsherniak, A.; Tamayo, P.; Santagata, S.; Ligon, A.; Greulich, H.; Berhoukim, R.; Amani, V.; Goumnerova, L.; Eberhart, C.G.; et al. Integrative genomic analysis of medulloblastoma identifies a molecular subgroup that drives poor clinical outcome. *J. Clin. Oncol.* **2011**, *29*, 1424–1430. [CrossRef] [PubMed]
13. Northcott, P.A.; Dubuc, A.M.; Pfister, S.; Taylor, M.D. Molecular subgroups of medulloblastoma. *Expert Rev. Neurother.* **2012**, *12*, 871–884. [CrossRef] [PubMed]
14. Northcott, P.A.; Jones, D.T.; Kool, M.; Robinson, G.W.; Gilbertson, R.J.; Cho, Y.J.; Pomeroy, S.L.; Korshunov, A.; Lichter, P.; Taylor, M.D.; et al. Medulloblastomics: The end of the beginning. *Nature Rev. Cancer* **2012**, *12*, 818–834. [CrossRef] [PubMed]
15. Taylor, M.D.; Northcott, P.A.; Korshunov, A.; Remke, M.; Cho, Y.J.; Clifford, S.C.; Eberhart, C.G.; Parsons, D.W.; Rutkowski, S.; Gajjar, A.; et al. Molecular subgroups of medulloblastoma: The current consensus. *Acta Neuropathol.* **2012**, *123*, 465–472. [CrossRef] [PubMed]
16. Northcott, P.A.; Buchhalter, I.; Morrissy, A.S.; Hovestadt, V.; Weischenfeldt, J.; Ehrenberger, T.; Grobner, S.; Segura-Wang, M.; Zichner, T.; Rudneva, V.A.; et al. The whole-genome landscape of medulloblastoma subtypes. *Nature* **2017**, *547*, 311–317. [CrossRef] [PubMed]
17. Gibson, P.; Tong, Y.; Robinson, G.; Thompson, M.C.; Currle, D.S.; Eden, C.; Kranenburg, T.A.; Hogg, T.; Poppleton, H.; Martin, J.; et al. Subtypes of medulloblastoma have distinct developmental origins. *Nature* **2010**, *468*, 1095–1099. [CrossRef] [PubMed]
18. Ellison, D.W.; Onilude, O.E.; Lindsey, J.C.; Lusher, M.E.; Weston, C.L.; Taylor, R.E.; Pearson, A.D.; Clifford, S.C. Beta-catenin status predicts a favorable outcome in childhood medulloblastoma: The united kingdom children's cancer study group brain tumour committee. *J. Clin. Oncol.* **2005**, *23*, 7951–7957. [CrossRef] [PubMed]
19. Northcott, P.A.; Korshunov, A.; Witt, H.; Hielscher, T.; Eberhart, C.G.; Mack, S.; Bouffet, E.; Clifford, S.C.; Hawkins, C.E.; French, P.; et al. Medulloblastoma comprises four distinct molecular variants. *J. Clin. Oncol.* **2011**, *29*, 1408–1414. [CrossRef] [PubMed]
20. Ramaswamy, V.; Remke, M.; Bouffet, E.; Bailey, S.; Clifford, S.C.; Doz, F.; Kool, M.; Dufour, C.; Vassal, G.; Milde, T.; et al. Risk stratification of childhood medulloblastoma in the molecular era: The current consensus. *Acta Neuropathol.* **2016**, *131*, 821–831. [CrossRef] [PubMed]
21. Zhukova, N.; Ramaswamy, V.; Remke, M.; Pfaff, E.; Shih, D.J.; Martin, D.C.; Castelo-Branco, P.; Baskin, B.; Ray, P.N.; Bouffet, E.; et al. Subgroup-specific prognostic implications of TP53 mutation in medulloblastoma. *J. Clin. Oncol.* **2013**, *31*, 2927–2935. [CrossRef] [PubMed]
22. Jones, D.T.; Jager, N.; Kool, M.; Zichner, T.; Hutter, B.; Sultan, M.; Cho, Y.J.; Pugh, T.J.; Hovestadt, V.; Stutz, A.M.; et al. Dissecting the genomic complexity underlying medulloblastoma. *Nature* **2012**, *488*, 100–105. [CrossRef] [PubMed]
23. Pugh, T.J.; Weeraratne, S.D.; Archer, T.C.; Pomeranz Krummel, D.A.; Auclair, D.; Bochicchio, J.; Carneiro, M.O.; Carter, S.L.; Cibulskis, K.; Erlich, R.L.; et al. Medulloblastoma exome sequencing uncovers subtype-specific somatic mutations. *Nature* **2012**, *488*, 106–110. [CrossRef] [PubMed]

24. Robinson, G.; Parker, M.; Kranenburg, T.A.; Lu, C.; Chen, X.; Ding, L.; Phoenix, T.N.; Hedlund, E.; Wei, L.; Zhu, X.; et al. Novel mutations target distinct subgroups of medulloblastoma. *Nature* **2012**, *488*, 43–48. [CrossRef] [PubMed]

25. Taylor, M.D.; Liu, L.; Raffel, C.; Hui, C.C.; Mainprize, T.G.; Zhang, X.; Agatep, R.; Chiappa, S.; Gao, L.; Lowrance, A.; et al. Mutations in sufu predispose to medulloblastoma. *Nature Genet.* **2002**, *31*, 306–310. [CrossRef] [PubMed]

26. Waszak, S.M.; Northcott, P.A.; Buchhalter, I.; Robinson, G.W.; Sutter, C.; Groebner, S.; Grund, K.B.; Brugieres, L.; Jones, D.T.W.; Pajtler, K.W.; et al. Spectrum and prevalence of genetic predisposition in medulloblastoma: A retrospective genetic study and prospective validation in a clinical trial cohort. *Lancet Oncol.* **2018**, *19*, 785–798. [CrossRef]

27. Pei, Y.; Moore, C.E.; Wang, J.; Tewari, A.K.; Eroshkin, A.; Cho, Y.J.; Witt, H.; Korshunov, A.; Read, T.A.; Sun, J.L.; et al. An animal model of MYC-driven medulloblastoma. *Cancer cell* **2012**, *21*, 155–167. [CrossRef] [PubMed]

28. Shih, D.J.; Northcott, P.A.; Remke, M.; Korshunov, A.; Ramaswamy, V.; Kool, M.; Luu, B.; Yao, Y.; Wang, X.; Dubuc, A.M.; et al. Cytogenetic prognostication within medulloblastoma subgroups. *J. Clin. Oncol.* **2014**, *32*, 886–896. [CrossRef] [PubMed]

29. Thompson, E.M.; Hielscher, T.; Bouffet, E.; Remke, M.; Luu, B.; Gururangan, S.; McLendon, R.E.; Bigner, D.D.; Lipp, E.S.; Perreault, S.; et al. Prognostic value of medulloblastoma extent of resection after accounting for molecular subgroup: A retrospective integrated clinical and molecular analysis. *Lancet Oncol* **2016**, *17*, 484–495. [CrossRef]

30. Wong, K.K.; Ragab, O.; Tran, H.N.; Pham, A.; All, S.; Waxer, J.; Olch, A.J. Acute toxicity of craniospinal irradiation with volumetric-modulated arc therapy in children with solid tumors. *Pediatr Blood Cancer* **2018**, *65*, e27050. [CrossRef] [PubMed]

31. Rutkowski, S.; Gerber, N.U.; von Hoff, K.; Gnekow, A.; Bode, U.; Graf, N.; Berthold, F.; Henze, G.; Wolff, J.E.; Warmuth-Metz, M.; et al. Treatment of early childhood medulloblastoma by postoperative chemotherapy and deferred radiotherapy. *Neuro-Oncology* **2009**, *11*, 201–210. [CrossRef] [PubMed]

32. Yock, T.I.; Yeap, B.Y.; Ebb, D.H.; Weyman, E.; Eaton, B.R.; Sherry, N.A.; Jones, R.M.; MacDonald, S.M.; Pulsifer, M.B.; Lavally, B.; et al. Long-term toxic effects of proton radiotherapy for paediatric medulloblastoma: A phase 2 single-arm study. *Lancet Oncol.* **2016**, *17*, 287–298. [CrossRef]

33. Ashley, D.M.; Merchant, T.E.; Strother, D.; Zhou, T.; Duffner, P.; Burger, P.C.; Miller, D.C.; Lyon, N.; Bonner, M.J.; Msall, M.; et al. Induction chemotherapy and conformal radiation therapy for very young children with nonmetastatic medulloblastoma: Children's oncology group study P9934. *J. Clin. Oncol.* **2012**, *30*, 3181–3186. [CrossRef] [PubMed]

34. Michalski, J.M.; Janss, A.; Vezina, G.; Gajjar, A.; Pollack, I.; Merchant, T.E.; FitzGerald, T.J.; Booth, T.; Tarbell, N.J.; Li, Y.; et al. Results of cog acns0331: A phase III trial of involved-field radiotherapy (IFRT) and low dose craniospinal irradiation (Ld-Csi) with chemotherapy in average-risk medulloblastoma: A report from the children's oncology group. *Int. J. Radiat. Oncol. Biol. Phys.* **2016**, *96*, 937–938. [CrossRef]

35. Johnson, S.B.; Hung, J.; Kapadia, N.; Oh, K.S.; Kim, M.; Hamstra, D.A. Spinal growth patterns following craniospinal irradiation in children with medulloblastoma. *Pract. Radiat. Oncol.* **2018**. [CrossRef] [PubMed]

36. Cohen, B.H.; Geyer, J.R.; Miller, D.C.; Curran, J.G.; Zhou, T.; Holmes, E.; Ingles, S.A.; Dunkel, I.J.; Hilden, J.; Packer, R.J.; et al. Pilot study of intensive chemotherapy with peripheral hematopoietic cell support for children less than 3 years of age with malignant brain tumors, the CCG-99703 phase I/II study. A report from the children's oncology group. *Pediat. Neurol.* **2015**, *53*, 31–46. [CrossRef] [PubMed]

37. St. Jude Children's Research Hospital; Genentech, Inc.; National Cancer Institute. A Clinical and Molecular Risk-Directed Therapy for Newly Diagnosed Medulloblastoma. Available online: https://clinicaltrials.gov/ct2/show/NCT01878617 (accessed on 21 September 2018).

38. Children's Oncology Group; National Cancer Institute. Reduced Craniospinal Radiation Therapy and Chemotherapy in Treating Younger Patients with Newly Diagnosed Wnt-Driven Medulloblastoma. Available online: https://clinicaltrials.gov/ct2/show/NCT02724579 (accessed on 21 September 2018).

39. Sidney Kimmel Comprehensive Cancer Center at Johns Hopkins. Study Assessing the Feasibility of a Surgery and Chemotherapy-Only in Children with Wnt Positive Medulloblastoma. Available online: https://clinicaltrials.gov/ct2/show/NCT02212574 (accessed on 21 September 2018).

40. LoRusso, P.M.; Rudin, C.M.; Reddy, J.C.; Tibes, R.; Weiss, G.J.; Borad, M.J.; Hann, C.L.; Brahmer, J.R.; Chang, I.; Darbonne, W.C.; et al. Phase I trial of hedgehog pathway inhibitor vismodegib (GDC-0449) in patients with refractory, locally advanced or metastatic solid tumors. *Clin. Cancer Res.* **2011**, *17*, 2502–2511. [CrossRef] [PubMed]

41. Romer, J.T.; Kimura, H.; Magdaleno, S.; Sasai, K.; Fuller, C.; Baines, H.; Connelly, M.; Stewart, C.F.; Gould, S.; Rubin, L.L.; et al. Suppression of the Shh pathway using a small molecule inhibitor eliminates medulloblastoma in Ptc1(+/-)p53(-/-) mice. *Cancer cell* **2004**, *6*, 229–240. [CrossRef] [PubMed]

42. Rohner, A.; Spilker, M.E.; Lam, J.L.; Pascual, B.; Bartkowski, D.; Li, Q.J.; Yang, A.H.; Stevens, G.; Xu, M.; Wells, P.A.; et al. Effective targeting of hedgehog signaling in a medulloblastoma model with PF-5274857, a potent and selective smoothened antagonist that penetrates the blood-brain barrier. *Mol. Cancer Ther.* **2012**, *11*, 57–65. [CrossRef] [PubMed]

43. Robinson, G.W.; Orr, B.A.; Wu, G.; Gururangan, S.; Lin, T.; Qaddoumi, I.; Packer, R.J.; Goldman, S.; Prados, M.D.; Desjardins, A.; et al. Vismodegib exerts targeted efficacy against recurrent sonic hedgehog-subgroup medulloblastoma: Results from phase II pediatric brain tumor consortium studies PBTC-025B and PBTC-032. *J. Clin. Oncol.* **2015**, *33*, 2646–2654. [CrossRef] [PubMed]

44. Kieran, M.W.; Chisholm, J.; Casanova, M.; Brandes, A.A.; Aerts, I.; Bouffet, E.; Bailey, S.; Leary, S.; MacDonald, T.J.; Mechinaud, F.; et al. Phase I study of oral sonidegib (LDE225) in pediatric brain and solid tumors and a phase II study in children and adults with relapsed medulloblastoma. *Neuro-Oncology* **2017**, *19*, 1542–1552. [CrossRef] [PubMed]

45. Buonamici, S.; Williams, J.; Morrissey, M.; Wang, A.; Guo, R.; Vattay, A.; Hsiao, K.; Yuan, J.; Green, J.; Ospina, B.; et al. Interfering with resistance to smoothened antagonists by inhibition of the PI3K pathway in medulloblastoma. *Sci. Transl. Med.* **2010**, *2*, 51ra70. [CrossRef] [PubMed]

46. Rudin, C.M.; Hann, C.L.; Laterra, J.; Yauch, R.L.; Callahan, C.A.; Fu, L.; Holcomb, T.; Stinson, J.; Gould, S.E.; Coleman, B.; et al. Treatment of medulloblastoma with hedgehog pathway inhibitor GDC-0449. *New Engl. J. Med.* **2009**, *361*, 1173–1178. [CrossRef] [PubMed]

47. Matheson, C.J.; Venkataraman, S.; Amani, V.; Harris, P.S.; Backos, D.S.; Donson, A.M.; Wempe, M.F.; Foreman, N.K.; Vibhakar, R.; Reigan, P. A WEE1 inhibitor analog of AZD1775 maintains synergy with cisplatin and demonstrates reduced single-agent cytotoxicity in medulloblastoma cells. *ACS Chem. Biol.* **2016**, *11*, 2066–2067. [CrossRef] [PubMed]

48. Hoffman, L.M.; Fouladi, M.; Olson, J.; Daryani, V.M.; Stewart, C.F.; Wetmore, C.; Kocak, M.; Onar-Thomas, A.; Wagner, L.; Gururangan, S.; et al. Phase I trial of weekly MK-0752 in children with refractory central nervous system malignancies: A pediatric brain tumor consortium study. *Child's nerv. Syst.* **2015**, *31*, 1283–1289. [CrossRef] [PubMed]

49. MacDonald, T.J.; Aguilera, D.; Castellino, R.C. The rationale for targeted therapies in medulloblastoma. *Neuro-Oncology* **2014**, *16*, 9–20. [CrossRef] [PubMed]

50. Henderson, J.J.; Wagner, J.P.; Hofmann, N.E.; Eide, C.A.; Cho, Y.J.; Druker, B.J.; Davare, M.A. Functional validation of the oncogenic cooperativity and targeting potential of tuberous sclerosis mutation in medullblastoma using a MYC-amplified model cell line. *Pediat. Blood Cancer* **2017**, *64*, e26553. [CrossRef] [PubMed]

51. National Cancer Institute. PI3K/mTOR Inhibitor LY3023414 in Treating Patients with Relapsed or Refractory Advanced Solid Tumors, Non-Hodgkin Lymphoma, or Histiocytic Disorders with TSC or PI3K/mTOR Mutations (a Pediatric Match Treatment Trial). Available online: https://clinicaltrials.gov/ct2/show/NCT03213678 (accessed on 21 September 2018).

52. Pei, Y.; Liu, K.W.; Wang, J.; Garancher, A.; Tao, R.; Esparza, L.A.; Maier, D.L.; Udaka, Y.T.; Murad, N.; Morrissy, S.; et al. HDAC and PI3K antagonists cooperate to inhibit growth of MYC-driven medulloblastoma. *Cancer cell* **2016**, *29*, 311–323. [CrossRef] [PubMed]

53. Bandopadhayay, P.; Bergthold, G.; Nguyen, B.; Schubert, S.; Gholamin, S.; Tang, Y.; Bolin, S.; Schumacher, S.E.; Zeid, R.; Masoud, S.; et al. Bet bromodomain inhibition of myc-amplified medulloblastoma. *Clin. Cancer Res.* **2014**, *20*, 912–925. [CrossRef] [PubMed]

54. National Cancer Institute. Vorinostat Combined with Isotretinoin and Chemotherapy in Treating Younger Patients with Embryonal Tumors of the Central Nervous System. Available online: https://clinicaltrials.gov/ct2/show/NCT00867178 (accessed on 21 September 2018).

55. Li, M.; Lockwood, W.; Zielenska, M.; Northcott, P.; Ra, Y.S.; Bouffet, E.; Yoshimoto, M.; Rutka, J.T.; Yan, H.; Taylor, M.D.; et al. Multiple *CDK/CYCLIND* genes are amplified in medulloblastoma and supratentorial primitive neuroectodermal brain tumor. *Cancer Genet.* **2012**, *205*, 220–231. [CrossRef] [PubMed]

56. Pediatric Brain Tumor Consortium; National Cancer Institute. Palbociclib Isethionate in Treating Younger Patients with Recurrent, Progressive, or Refractory Central Nervous System Tumors. Available online: https://clinicaltrials.gov/ct2/show/NCT02255461 (accessed on 21 September 2018).

57. Fouladi, M.; Stewart, C.F.; Blaney, S.M.; Onar-Thomas, A.; Schaiquevich, P.; Packer, R.J.; Goldman, S.; Geyer, J.R.; Gajjar, A.; Kun, L.E.; et al. A molecular biology and phase II trial of lapatinib in children with refractory CNS malignancies: A pediatric brain tumor consortium study. *J. Neuro-oncol.* **2013**, *114*, 173–179. [CrossRef] [PubMed]

58. Jakacki, R.I.; Hamilton, M.; Gilbertson, R.J.; Blaney, S.M.; Tersak, J.; Krailo, M.D.; Ingle, A.M.; Voss, S.D.; Dancey, J.E.; Adamson, P.C. Pediatric phase I and pharmacokinetic study of erlotinib followed by the combination of erlotinib and temozolomide: A children's oncology group phase I consortium study. *J. Clin. Oncol.* **2008**, *26*, 4921–4927. [CrossRef] [PubMed]

59. Kieran, M.W.; Chi, S.; Goldman, S.; Onar-Thomas, A.; Poussaint, T.Y.; Vajapeyam, S.; Fahey, F.; Wu, S.; Turner, D.C.; Stewart, C.F.; et al. A phase I trial and pk study of cediranib (AZD2171), an orally bioavailable pan-VEGFR inhibitor, in children with recurrent or refractory primary CNS tumors. *Child's Nerv. Syst.* **2015**, *31*, 1433–1445. [CrossRef] [PubMed]

60. Piha-Paul, S.A.; Shin, S.J.; Vats, T.; Guha-Thakurta, N.; Aaron, J.; Rytting, M.; Kleinerman, E.; Kurzrock, R. Pediatric patients with refractory central nervous system tumors: Experiences of a clinical trial combining bevacizumab and temsirolimus. *Anticancer Res.* **2014**, *34*, 1939–1945. [PubMed]

61. St. Jude Children's Research Hospital; Novartis Pharmaceuticals. SJDAWN: St. Jude Children's Research Hospital Phase 1 Study Evaluating Molecularly-Driven Doublet Therapies for Children and Young Adults with Recurrent Brain Tumors. Available online: https://clinicaltrials.gov/ct2/show/NCT03434262 (accessed on 21 September 2018).

62. Martin, A.M.; Nirschl, C.J.; Polanczyk, M.J.; Bell, W.R.; Nirschle, T.R.; Harris-Bookman, S.; Phallen, J.; Hicks, J.; Martinez, D.; Ogurtsova, A.; et al. PD-L1 expression in medulloblastoma: An evaluation by subgroup. *Oncotarget* **2018**, *9*, 19177–19191. [CrossRef] [PubMed]

63. Pham, C.D.; Mitchell, D.A. Know your neighbors: Different tumor microenvironments have implications in immunotherapeutic targeting strategies across MB subgroups. *Oncoimmunology* **2016**, *5*, e1144002. [CrossRef] [PubMed]

64. Pham, C.D.; Flores, C.; Yang, C.; Pinheiro, E.M.; Yearley, J.H.; Sayour, E.J.; Pei, Y.; Moore, C.; McLendon, R.E.; Huang, J.; et al. Differential immune microenvironements and response to immune checkpoint blockade among molecular subtypes of murine medulloblastoma. *Clin. Cancer Res.* **2016**, *22*, 582–595. [CrossRef] [PubMed]

65. Schramm, A.; Lode, H. MYCN-targeting vaccines and immunotherapeutics. *Hum. Vaccines Immunother.* **2016**, *12*, 2257–2258. [CrossRef] [PubMed]

66. Lal, S.; Carrera, D.; Phillips, J.J.; Weiss, W.A.; Raffel, C. An oncolytic measles virus-sensitive group 3 medulloblastoma model in immune-competent mice. *Neuro-Oncology* **2018**. [CrossRef] [PubMed]

67. Rorke, L.B.; Packer, R.J.; Biegel, J.A. Central nervous system atypical teratoid/rhabdoid tumors of infancy and childhood: Definition of an entity. *J. Neurosurg.* **1996**, *85*, 56–65. [CrossRef] [PubMed]

68. McGovern, S.L.; Grosshans, D.; Mahajan, A. Embryonal brain tumors. *Cancer J.* **2014**, *20*, 397–402. [CrossRef] [PubMed]

69. Wetmore, C.; Boyett, J.; Li, S.; Lin, T.; Bendel, A.; Gajjar, A.; Orr, B.A. Alisertib is active as single agent in recurrent atypical teratoid rhabdoid tumors in 4 children. *Neuro-Oncology* **2015**, *17*, 882–888. [CrossRef] [PubMed]

70. Fruhwald, M.C.; Biegel, J.A.; Bourdeaut, F.; Roberts, C.W.; Chi, S.N. Atypical teratoid/rhabdoid tumors-current concepts, advances in biology, and potential future therapies. *Neuro-Oncology* **2016**, *18*, 764–778. [CrossRef] [PubMed]

71. Jin, B.; Feng, X.Y. MRI features of atypical teratoid/rhabdoid tumors in children. *Pediatr. Radiol.* **2013**, *43*, 1001–1008. [CrossRef] [PubMed]

72. Warmuth-Metz, M.; Bison, B.; Dannemann-Stern, E.; Kortmann, R.; Rutkowski, S.; Pietsch, T. Ct and mr imaging in atypical teratoid/rhabdoid tumors of the central nervous system. *Neuroradiology* **2008**, *50*, 447–452. [CrossRef] [PubMed]

73. Nowak, J.; Nemes, K.; Hohm, A.; Vandergrift, L.A.; Hasselblatt, M.; Johann, P.D.; Kool, M.; Fruhwald, M.C.; Warmuth-Metz, M. Magnetic resonance imaging surrogates of molecular subgroups in atypical teratoid/rhabdoid tumor (ATRT). *Neuro-Oncology* **2018**. [CrossRef] [PubMed]

74. Bikowska, B.; Grajkowska, W.; Jozwiak, J. Atypical teratoid/rhabdoid tumor: Short clinical description and insight into possible mechanism of the disease. *Eur. J. Neurol.* **2011**, *18*, 813–818. [CrossRef] [PubMed]

75. Biegel, J.A. Molecular genetics of atypical teratoid/rhabdoid tumor. *Neurosurg. Focus* **2006**, *20*, 1–7. [CrossRef]

76. Haberler, C.; Laggner, U.; Slavc, I.; Czech, T.; Ambros, I.M.; Ambros, P.F.; Budka, H.; Hainfellner, J.A. Immunohistochemical analysis of INI1 protein in malignant pediatric CNS tumors: Lack of INI1 in atypical teratoid/rhabdoid tumors and in a fraction of primitive neuroectodermal tumors without rhabdoid phenotype. *Am. J. Surg Pathol.* **2006**, *30*, 1462–1468. [CrossRef] [PubMed]

77. Eaton, K.W.; Tooke, L.S.; Wainwright, L.M.; Judkins, A.R.; Biegel, J.A. Spectrum of *smarcb1/ini1* mutations in familial and sporadic rhabdoid tumors. *Pediatr. Blood Cancer* **2011**, *56*, 7–15. [CrossRef] [PubMed]

78. Biegel, J.A.; Zhou, J.Y.; Rorke, L.B.; Stenstrom, C.; Wainwright, L.M.; Fogelgren, B. Germ-line and acquired mutations of *ini1* in atypical teratoid and rhabdoid tumors. *Cancer Res.* **1999**, *59*, 74–79. [PubMed]

79. Torchia, J.; Picard, D.; Lafay-Cousin, L.; Hawkins, C.E.; Kim, S.K.; Letourneau, L.; Ra, Y.S.; Ho, K.C.; Chan, T.S.; Sin-Chan, P.; et al. Molecular subgroups of atypical teratoid rhabdoid tumours in children: An integrated genomic and clinicopathological analysis. *Lancet Oncol.* **2015**, *16*, 569–582. [CrossRef]

80. Hasselblatt, M.; Gesk, S.; Oyen, F.; Rossi, S.; Viscardi, E.; Giangaspero, F.; Giannini, C.; Judkins, A.R.; Fruhwald, M.C.; Obser, T.; et al. Nonsense mutation and inactivation of *smarca4* (*brg1*) in an atypical teratoid/rhabdoid tumor showing retained *smarcb1* (*ini1*) expression. *Am. J. Surg Pathol* **2011**, *35*, 933–935. [CrossRef] [PubMed]

81. Birks, D.K.; Donson, A.M.; Patel, P.R.; Dunham, C.; Muscat, A.; Algar, E.M.; Ashley, D.M.; Kleinschmidt-Demasters, B.K.; Vibhakar, R.; Handler, M.H.; et al. High expression of bmp pathway genes distinguishes a subset of atypical teratoid/rhabdoid tumors associated with shorter survival. *Neuro-Oncology* **2011**, *13*, 1296–1307. [CrossRef] [PubMed]

82. Johann, P.D.; Erkek, S.; Zapatka, M.; Kerl, K.; Buchhalter, I.; Hovestadt, V.; Jones, D.T.W.; Sturm, D.; Hermann, C.; Segura Wang, M.; et al. Atypical teratoid/rhabdoid tumors are comprised of three epigenetic subgroups with distinct enhancer landscapes. *Cancer Cell.* **2016**, *29*, 379–393. [CrossRef] [PubMed]

83. Han, Z.Y.; Richer, W.; Freneaux, P.; Chauvin, C.; Lucchesi, C.; Guillemot, D.; Grison, C.; Lequin, D.; Pierron, G.; Masliah-Planchon, J.; et al. The occurrence of intracranial rhabdoid tumours in mice depends on temporal control of smarcb1 inactivation. *Nat. Commun.* **2016**, *7*, 10421. [CrossRef] [PubMed]

84. Ginn, K.F.; Gajjar, A. Atypical teratoid rhabdoid tumor: Current therapy and future directions. *Front. Oncol.* **2012**, *2*, 114. [CrossRef] [PubMed]

85. Chi, S.N.; Zimmerman, M.A.; Yao, X.; Cohen, K.J.; Burger, P.; Biegel, J.A.; Rorke-Adams, L.B.; Fisher, M.J.; Janss, A.; Mazewski, C.; et al. Intensive multimodality treatment for children with newly diagnosed cns atypical teratoid rhabdoid tumor. *J. Clin. Oncol.* **2009**, *27*, 385–389. [CrossRef] [PubMed]

86. Tekautz, T.M.; Fuller, C.E.; Blaney, S.; Fouladi, M.; Broniscer, A.; Merchant, T.E.; Krasin, M.; Dalton, J.; Hale, G.; Kun, L.E.; et al. Atypical teratoid/rhabdoid tumors (ATRT): Improved survival in children 3 years of age and older with radiation therapy and high-dose alkylator-based chemotherapy. *J. Clin. Oncol.* **2005**, *23*, 1491–1499. [CrossRef] [PubMed]

87. Fangusaro, J.; Finlay, J.; Sposto, R.; Ji, L.; Saly, M.; Zacharoulis, S.; Asgharzadeh, S.; Abromowitch, M.; Olshefski, R.; Halpern, S.; et al. Intensive chemotherapy followed by consolidative myeloablative chemotherapy with autologous hematopoietic cell rescue (AuHCR) in young children with newly diagnosed supratentorial primitive neuroectodermal tumors (sPNETs): Report of the head start I and II experience. *Pediatr. Blood Cancer* **2008**, *50*, 312–318. [PubMed]

88. Lafay-Cousin, L.; Hawkins, C.; Carret, A.S.; Johnston, D.; Zelcer, S.; Wilson, B.; Jabado, N.; Scheinemann, K.; Eisenstat, D.; Fryer, C.; et al. Central nervous system atypical teratoid rhabdoid tumours: The canadian paediatric brain tumour consortium experience. *Eur J. Cancer* **2012**, *48*, 353–359. [CrossRef] [PubMed]

89. Hilden, J.M.; Meerbaum, S.; Burger, P.; Finlay, J.; Janss, A.; Scheithauer, B.W.; Walter, A.W.; Rorke, L.B.; Biegel, J.A. Central nervous system atypical teratoid/rhabdoid tumor: Results of therapy in children enrolled in a registry. *J. Clin. Oncol.* **2004**, *22*, 2877–2884. [CrossRef] [PubMed]
90. Athale, U.H.; Duckworth, J.; Odame, I.; Barr, R. Childhood atypical teratoid rhabdoid tumor of the central nervous system: A meta-analysis of observational studies. *J. Pediatr. Hematol. Oncol.* **2009**, *31*, 651–663. [CrossRef] [PubMed]
91. Buscariollo, D.L.; Park, H.S.; Roberts, K.B.; Yu, J.B. Survival outcomes in atypical teratoid rhabdoid tumor for patients undergoing radiotherapy in a surveillance, epidemiology, and end results analysis. *Cancer* **2012**, *118*, 4212–4219. [CrossRef] [PubMed]
92. Kurmasheva, R.T.; Sammons, M.; Favours, E.; Wu, J.; Kurmashev, D.; Cosmopoulos, K.; Keilhack, H.; Klaus, C.R.; Houghton, P.J.; Smith, M.A. Initial testing (stage 1) of tazemetostat (EPZ-6438), a novel EZH2 inhibitor, by the pediatric preclinical testing program. *Pediatr. Blood Cancer* **2017**, *64*, e26218. [CrossRef] [PubMed]
93. Unland, R.; Borchardt, C.; Clemens, D.; Kool, M.; Dirksen, U.; Fruhwald, M.C. Analysis of the antiproliferative effects of 3-deazaneoplanocin a in combination with standard anticancer agents in rhabdoid tumor cell lines. *Anticancer Drugs* **2015**, *26*, 301–311. [CrossRef] [PubMed]
94. Ribrag, V.; Soria, J.C.; Reyderman, L.; Chen, R.; Salazar, P.; Kumar, N.; Kuznetsov, G.; Keilhack, H.; Ottesen, L.H.; Italiano, A. O7.2phase 1 first-in-human study of the enhancer of zeste-homolog 2 (EZH2) histone methyl transferase inhibitor E7438. *Annals of Oncology* **2015**, *26*. [CrossRef]
95. Kerl, K.; Holsten, T.; Fruhwald, M.C. Rhabdoid tumors: Clinical approaches and molecular targets for innovative therapy. *Pediatr. Hematol. Oncol.* **2013**, *30*, 587–604. [CrossRef] [PubMed]
96. Tang, Y.; Gholamin, S.; Schubert, S.; Willardson, M.I.; Lee, A.; Bandopadhayay, P.; Bergthold, G.; Masoud, S.; Nguyen, B.; Vue, N.; et al. Epigenetic targeting of hedgehog pathway transcriptional output through bet bromodomain inhibition. *Nat. Med.* **2014**, *20*, 732–740. [CrossRef] [PubMed]
97. Knipstein, J.A.; Birks, D.K.; Donson, A.M.; Alimova, I.; Foreman, N.K.; Vibhakar, R. Histone deacetylase inhibition decreases proliferation and potentiates the effect of ionizing radiation in atypical teratoid/rhabdoid tumor cells. *Neuro-Oncology* **2012**, *14*, 175–183. [CrossRef] [PubMed]
98. Tsikitis, M.; Zhang, Z.; Edelman, W.; Zagzag, D.; Kalpana, G.V. Genetic ablation of cyclin D1 abrogates genesis of rhabdoid tumors resulting from Ini1 loss. *Proc. Natl. Acad. Sci. USA* **2005**, *102*, 12129–12134. [CrossRef] [PubMed]
99. Venneti, S.; Le, P.; Martinez, D.; Eaton, K.W.; Shyam, N.; Jordan-Sciutto, K.L.; Pawel, B.; Biegel, J.A.; Judkins, A.R. P16^{INK4a} and p14^{ARF} tumor suppressor pathways are deregulated in malignant rhabdoid tumors. *J. Neuropathol. Exp. Neurol.* **2011**, *70*, 596–609. [CrossRef] [PubMed]
100. Hashizume, R.; Zhang, A.; Mueller, S.; Prados, M.D.; Lulla, R.R.; Goldman, S.; Saratsis, A.M.; Mazar, A.P.; Stegh, A.H.; Cheng, S.Y.; et al. Inhibition of DNA damage repair by the CDK4/6 inhibitor palbociclib delays irradiated intracranial atypical teratoid rhabdoid tumor and glioblastoma xenograft regrowth. *Neuro-Oncology* **2016**, *18*, 1519–1528. [CrossRef] [PubMed]
101. Geoerger, B.; Bourdeaut, F.; DuBois, S.G.; Fischer, M.; Geller, J.I.; Gottardo, N.G.; Marabelle, A.; Pearson, A.D.J.; Modak, S.; Cash, T.; et al. A phase I study of the CDK4/6 inhibitor ribociclib (LEE011) in pediatric patients with malignant rhabdoid tumors, neuroblastoma, and other solid tumors. *Clin. Cancer Res.* **2017**, *23*, 2433–2441. [CrossRef] [PubMed]
102. Pfister, S.; Remke, M.; Castoldi, M.; Bai, A.H.; Muckenthaler, M.U.; Kulozik, A.; von Deimling, A.; Pscherer, A.; Lichter, P.; Korshunov, A. Novel genomic amplification targeting the microrna cluster at 19q13.42 in a pediatric embryonal tumor with abundant neuropil and true rosettes. *Acta Neuropathol.* **2009**, *117*, 457–464. [CrossRef] [PubMed]
103. Korshunov, A.; Sturm, D.; Ryzhova, M.; Hovestadt, V.; Gessi, M.; Jones, D.T.; Remke, M.; Northcott, P.; Perry, A.; Picard, D.; et al. Embryonal tumor with abundant neuropil and true rosettes (ETANTR), ependymoblastoma, and medulloepithelioma share molecular similarity and comprise a single clinicopathological entity. *Acta Neuropathol.* **2014**, *128*, 279–289. [CrossRef] [PubMed]
104. Spence, T.; Perotti, C.; Sin-Chan, P.; Picard, D.; Wu, W.; Singh, A.; Anderson, C.; Blough, M.D.; Cairncross, J.G.; Lafay-Cousin, L.; et al. A novel C19MC amplified cell line links Lin28/let-7 to mtor signaling in embryonal tumor with multilayered rosettes. *Neuro-Oncology* **2014**, *16*, 62–71. [CrossRef] [PubMed]

105. Horwitz, M.; Dufour, C.; Leblond, P.; Bourdeaut, F.; Faure-Conter, C.; Bertozzi, A.I.; Delisle, M.B.; Palenzuela, G.; Jouvet, A.; Scavarda, D.; et al. Embryonal tumors with multilayered rosettes in children: The sfce experience. *Childs Nerv Syst* **2016**, *32*, 299–305. [CrossRef] [PubMed]

106. Wang, B.; Gogia, B.; Fuller, G.N.; Ketonen, L.M. Embryonal tumor with multilayered rosettes, c19mc-altered: Clinical, pathological, and neuroimaging findings. *J. Neuroimaging* **2018**, *28*, 483–489. [CrossRef] [PubMed]

107. Ceccom, J.; Bourdeaut, F.; Loukh, N.; Rigau, V.; Milin, S.; Takin, R.; Richer, W.; Uro-Coste, E.; Couturier, J.; Bertozzi, A.I.; et al. Embryonal tumor with multilayered rosettes: Diagnostic tools update and review of the literature. *Clin. Neuropathol.* **2014**, *33*, 15–22. [CrossRef] [PubMed]

108. Picard, D.; Miller, S.; Hawkins, C.E.; Bouffet, E.; Rogers, H.A.; Chan, T.S.; Kim, S.K.; Ra, Y.S.; Fangusaro, J.; Korshunov, A.; et al. Markers of survival and metastatic potential in childhood CNS primitive neuro-ectodermal brain tumours: An integrative genomic analysis. *Lancet Oncol.* **2012**, *13*, 838–848. [CrossRef]

109. Alexiou, G.A.; Stefanaki, K.; Vartholomatos, G.; Sfakianos, G.; Prodromou, N.; Moschovi, M. Embryonal tumor with abundant neuropil and true rosettes: A systematic literature review and report of 2 new cases. *J. Child Neurol.* **2013**, *28*, 1709–1715. [CrossRef] [PubMed]

110. Kleinman, C.L.; Gerges, N.; Papillon-Cavanagh, S.; Sin-Chan, P.; Pramatarova, A.; Quang, D.A.; Adoue, V.; Busche, S.; Caron, M.; Djambazian, H.; et al. Fusion of *TTYH1* with the C19MC microRNA cluster drives expression of a brain-specific *DNMT3B* isoform in the embryonal brain tumor ETMR. *Nat. Genet.* **2014**, *46*, 39–44. [CrossRef] [PubMed]

111. Braasch, I.; Bobe, J.; Guiguen, Y.; Postlethwait, J.H. Reply to: 'subfunctionalization versus neofunctionalization after whole-genome duplication'. *Nat. Genet.* **2018**, *50*, 910–911. [CrossRef] [PubMed]

112. Neumann, J.E.; Wefers, A.K.; Lambo, S.; Bianchi, E.; Bockstaller, M.; Dorostkar, M.M.; Meister, V.; Schindler, P.; Korshunov, A.; von Hoff, K.; et al. A mouse model for embryonal tumors with multilayered rosettes uncovers the therapeutic potential of sonic-hedgehog inhibitors. *Nat. Med.* **2017**, *23*, 1191–1202. [CrossRef] [PubMed]

113. Schmidt, C.; Schubert, N.A.; Brabetz, S.; Mack, N.; Schwalm, B.; Chan, J.A.; Selt, F.; Herold-Mende, C.; Witt, O.; Milde, T.; et al. Preclinical drug screen reveals topotecan, actinomycin D, and volasertib as potential new therapeutic candidates for ETMR brain tumor patients. *Neuro-Oncology* **2017**, *19*, 1607–1617. [CrossRef] [PubMed]

© 2018 by the authors. Licensee MDPI, Basel, Switzerland. This article is an open access article distributed under the terms and conditions of the Creative Commons Attribution (CC BY) license (http://creativecommons.org/licenses/by/4.0/).

bioengineering

MDPI

Review

Neurocognitive and Psychosocial Outcomes in Pediatric Brain Tumor Survivors

Peter L. Stavinoha [1,*]**, Martha A. Askins** [1]**, Stephanie K. Powell** [2]**, Natasha Pillay Smiley** [2] **and Rhonda S. Robert** [1]

1   The University of Texas MD Anderson Cancer Center, Houston, TX 77030, USA;
    maskins@mdanderson.org (M.A.A.); rrobert@mdanderson.org (R.S.R.)
2   Ann and Robert H. Lurie Children's Hospital of Chicago and Northwestern Feinberg School of Medicine,
    Chicago, IL 60611, USA; SkPowell@luriechildrens.org (S.K.P.); NPillaySmiley@luriechildrens.org (N.P.S.)
*   Correspondence: pstavinoha@mdanderson.org; Tel.: +1-713-794-4066

Received: 3 August 2018; Accepted: 8 September 2018; Published: 11 September 2018

Abstract: The late neurocognitive and psychosocial effects of treatment for pediatric brain tumor (PBT) represent important areas of clinical focus and ongoing research. Neurocognitive sequelae and associated problems with learning and socioemotional development negatively impact PBT survivors' overall health-related quality of life, educational attainment and employment rates. Multiple factors including tumor features and associated complications, treatment methods, individual protective and vulnerability factors and accessibility of environmental supports contribute to the neurocognitive and psychosocial outcomes in PBT survivors. Declines in overall measured intelligence are common and may persist years after treatment. Core deficits in attention, processing speed and working memory are postulated to underlie problems with overall intellectual development, academic achievement and career attainment. Additionally, psychological problems after PBT can include depression, anxiety and psychosocial adjustment issues. Several intervention paradigms are briefly described, though to date research on innovative, specific and effective interventions for neurocognitive late effects is still in its early stages. This article reviews the existing research for understanding PBT late effects and highlights the need for innovative research to enhance neurocognitive and psychosocial outcomes in PBT survivors.

Keywords: pediatric brain tumor; late effects; neurocognitive; cognitive; psychosocial; survivorship

---

1. Introduction

Approximately 2970 children and 1170 adolescents are diagnosed with brain and central nervous system tumors in the United States annually [1]. As survival rates following pediatric brain tumor (PBT) increase with improvements in detection and intervention, focus has increased on monitoring and managing the late effects of both disease and treatment (i.e., delayed emergence of neurocognitive, emotional and socioemotional sequelae). It is estimated that as many as 40 to 100% of survivors demonstrate impairment in at least one neurocognitive domain [2] and adult survivors of PBT report the poorest health-related quality of life among all childhood cancers [3]. Neurocognitive and psychosocial late effects are associated with lower high school and college graduation rates and increased likelihood of unemployment [4–9], all of which may adversely impact quality of life.

Late effects of treatment for PBT typically emerge in the first few years following treatment and clinically may range from mild performance difficulties that are easily accommodated to severe deficits in functioning that result in the ongoing need for support into adulthood. Here we provide a broad overview of neurocognitive and psychosocial late effects of treatment for PBT, including discussion of significant risk factors and pathophysiology of late effects, a summary of intervention paradigms and discussion of future opportunities to improve outcomes for survivors.

2. Factors Related to Expression of Late Effects

To understand the mechanisms of late effects, a host of factors must be considered. It is important to recognize that isolating the influence of any one variable among the multiple confounding and interacting variables is a consistent challenge in late effects research. With that caveat, factors with documented relationships to outcomes include tumor variables, treatment paradigms and potential moderating variables related to individual patient characteristics and environmental factors.

2.1. Tumor Variables

Tumor size has been associated with lower overall intelligence [1]. Higher risk pathology, such as medulloblastoma, has been associated with poorer neurocognitive outcomes, evident on measures of intelligence, aspects of attention, working memory and processing speed [2,3]. Tumor location is integral, in part due to associated complications. For example, 70–80% of children with posterior fossa tumors develop obstructive hydrocephalus, with approximately 30% requiring cerebrospinal fluid diversion via ventriculoperitoneal shunt or endoscopic third ventriculostomy [4]. Hydrocephalus has been shown to independently correlate with neurocognitive deficits even with otherwise uniform chemotherapy and radiation treatment [5] and is associated with poorer long term intellectual outcomes, regardless of tumor type [6]. Some evidence suggests that children with infratentorial tumors have greater neurocognitive burden than those with supratentorial tumors [7].

2.2. Treatment Variables

Advances in neurosurgical techniques over the last few decades have led to improved histologic diagnosis and decreased morbidity and mortality. Some tumors require only neurosurgical intervention. Still, studies suggest at least some short-term risk for neurocognitive deficits within the first year post surgery [8–10]. For example, even with refined neurosurgical practice, the post-surgical complication of posterior fossa syndrome (also known as cerebellar mutism) still occurs in up to 31% of children with infratentorial tumors [11]. This poorly understood entity has been attributed to disruption of cerebello-thalamo-cerebral pathways and is characterized by a unique constellation of symptoms that emerge approximately 24–48 h after surgery, including diminished speech, ataxia, emotional/behavioral lability and apathy. Although the speech and neurologic sequelae often improve with time and rehabilitation, recent evidence suggests worse overall neurocognitive outcomes for PBT survivors who experienced posterior fossa syndrome relative to those who did not [12].

Cranial radiation therapy (CRT) is often considered the most significant treatment-related risk factor for development of neurocognitive late effects [13]. CRT has been associated with significant declines in multiple neurocognitive domains that may continue for years post treatment [14]. Changes to white matter have received much attention as a mechanism of neurocognitive decline following radiation therapy, including decreased normal appearing white matter [15–17]. Cranial radiation also affects the growth of new neurons in the hippocampus [18] and decreased hippocampal volume has been associated with specific memory deficits [19] in PBT survivors. Further, working memory performance has been specifically associated with white matter integrity within cerebello-thalamo-cerebral pathways [20].

Effects of chemotherapy alone are difficult to isolate in the context of other treatment paradigms such as surgery and CRT, as well as in the presence of other tumor related variables and complications as reviewed above. While chemotherapy is thought to be less toxic than radiation therapy, specific chemotherapy agents are known to carry direct risk for cognitive impairment [21–23] as well as an indirect risk related to ototoxicity [24,25]. Further, concomitant chemotherapy and radiation appears to result in greater cognitive and educational burden compared to CRT alone [26,27].

2.3. Individual Patient and Environmental Characteristics

Age at diagnosis and treatment, as well as time since treatment, moderate neurocognitive outcomes in PBT survivors. Specifically, younger age at diagnosis and treatment has been associated with lower intellectual ability, processing speed, working memory, aspects of attention and academic performance [2,3,28]. In fact, a recent meta-analysis found time since treatment more predictive of overall intelligence than treatment modality in PBT survivors [13]. Further, higher levels of cognitive ability prior to treatment have been associated with greater declines in functioning after PBT treatment [2,28].

Additionally, environmental factors including low socioeconomic status and high stress levels may increase risk for poor neurocognitive and psychosocial outcomes [29–33]. It is well established that survivors of childhood cancer miss a considerable amount of school even after treatment is complete [34], though there is a paucity of research investigating the impact of this and other patient-specific experiential factors on neurocognitive outcome. There have been mixed research findings regarding the impact of gender on cognitive outcomes. While some studies have suggested female medulloblastoma survivors are at higher risk for poorer neurocognitive outcomes than males [28,35,36], others have failed to replicate this finding [5,37,38].

Multiple variables contribute to—and moderate—neurocognitive outcomes in PBT survivors as depicted in Figure 1 adapted from Dennis [39] and Baum, et al. [40]. Moreover, with increasing survivorship and risk-mitigating modifications to treatment regimens, it has become increasingly important to assess neurocognitive and psychological outcomes on an individual level. Several organizations have published psychosocial standards of care for long-term PBT survivors, including the Children's Oncology Group [41] and the Psychosocial Standards of Care Project for Childhood Cancer [42]. Proposed clinical services range from clinical surveillance to comprehensive neuropsychological evaluations [40].

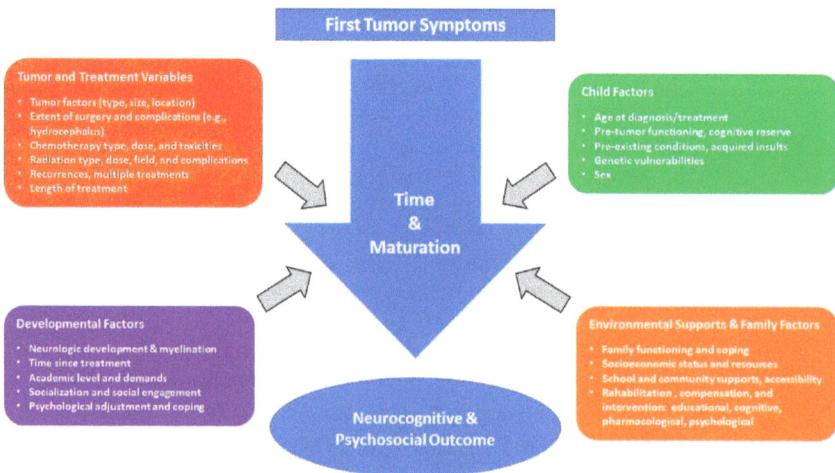

Figure 1. Factors affecting outcome in PBT survivors.

3. Late Effects of PBT

3.1. Neurocognitive Outcomes

Early studies of neurocognitive outcomes for survivors of PBT focused on global cognitive dysfunction, typically investigating the impact of brain tumors and their treatment on IQ scores and their trajectory over time. More recently, studies have identified specific cognitive functions at greatest risk, believed to represent core deficits that contribute to broader difficulties.

3.1.1. Intellectual Functioning

Declines in IQ are evident in PBT survivors as early as the first year following diagnosis and treatment [43], with potential gradual progression over the next 5 to 7 years [44–46]. Few studies have followed survivors longer; however, a handful of studies exploring adult survivors have reported IQs approximately one full standard deviation below healthy controls or the population mean [47–49]. Overall intelligence, as well as language-based abilities and non-language abilities, have all been shown to be impacted, with greatest effect sizes for nonverbal functions [13,48,50]. This may be related to the demands commonly administered nonverbal tasks place on visual attention, spatial processing and timed performance [13].

Notably, the decline in IQ scores evident in PBT survivors is related to a failure to make age-appropriate gains over time rather than an actual loss of skill. This was demonstrated in a seminal study [51] in which survivors of medulloblastoma achieved gains in raw scores but only at 49% to 62% the rate of their healthy same-age peers.

The impact of pediatric brain tumors on IQ appears mediated by age, disease and treatment variables. Children diagnosed and treated at a young age (<7) are at greatest risk [52,53], with a potentially more rapid initial decline that plateaus compared to older children, who display a slower, more protracted course [14,52]. Multiple studies have established CRT to carry substantial risk to IQ, mediated by dose, delivery and target [44,45,54,55]. Proton beam radiation (PBRT) has been proposed to carry less neurocognitive risk relative to traditional CRT [56]. Preliminary evidence suggests potential sparing of cognitive and academic functions [57], particularly with focal PBRT [58,59].

3.1.2. Core Deficits—Attention, Processing Speed, & Working Memory

Problems with attention, working memory and processing speed are some of the most commonly reported findings in studies of neurocognitive late effects [2,46,48,60,61]. In a recent meta-analysis, Robinson et al. [50] reported medium to large negative effect sizes for survivors of posterior fossa tumors in multiple cognitive domains, with the largest in attention (Hedges' $g = -1.69$) and processing speed ($g = -1.40$). Moreover, numerous studies have demonstrated a pattern of declining processing speed and working memory scores over time [2,14,45,51,62].

Notably, slowed processing speed is very common in long-term PBT survivors, regardless of tumor type [62]. Processing speed is also suggested to be the most significantly impacted cognitive domain subsequent to treatment for medulloblastoma [2]. Within developmental models, processing speed is conceptualized as a foundational capacity upon which other more complex cognitive abilities are dependent [49,63,64]. For example, analyses have demonstrated age-related gains in processing speed to account for the vast majority of age-related gains in working memory [64]; in addition, these tandem gains in processing speed and working memory occur with a corresponding improvement in intellectual ability.

Given this, it has been proposed that treatment-related deficits in processing speed, attention and working memory are the driving force behind the slowed rate of cognitive development and academic achievement observed in PBT survivors [2,31,49,51,65,66]. Some work has been done developing and evaluating specific interventions targeting these core deficits, although this remains an area of need for continued investigation.

3.1.3. Other Cognitive Functions

Meta-analyses of cognitive late effects in pediatric brain tumor survivors [48,50] have revealed large effect sizes in visually-based tasks, including nonverbal IQ and visual-spatial processing (Hedges' $g = -0.88$ to -1.29), as well as medium effect sizes in visual memory ($g = -0.68$). Studies in other cancers suggest exposure to cranial radiation carries risk for visual-spatial deficits [63]. However, a recent study looking explicitly at children with cerebellar low-grade gliomas requiring surgery alone suggests cerebellar involvement may be sufficient to cause visual-spatial impairment [67].

Executive functions refer to cognitive processes necessary for self-regulation and self-management of thinking, emotions and behavior, ranging from basic attentional and inhibitory control to more complex cognitive flexibility, set shifting and planning. Related deficits have been shown in PBT survivors relative to typically developing peers, both in terms of performance on clinical measures [68,69] and per standardized parent report [70,71]. Interestingly, PBT survivors may have limited awareness of their deficits, characterized by poor metacognition and unrealistic expectations of their abilities [46]. Executive deficits have been specifically associated with poorer long-term outcomes including lower rates of high school graduation and full employment [47,72].

Although less studied than other cognitive domains, memory problems have been demonstrated in PBT survivors across tumor type [73–78]. The majority of children with medulloblastoma demonstrate some level of memory impairment; survivors of astrocytoma may be comparatively less impaired but still perform below normal control groups [74,75]. Memory difficulties persist at least into adolescence and young adulthood following PBT [79]. While verbal memory appears more impaired than visual memory [48,50], longitudinal progressive decline has been observed in visual but not verbal memory [14,80].

Language is another area of known risk following treatment for PBT that has not been extensively researched. Meta-analyses have reported medium to large negative effect sizes in PBT survivors, in both general language abilities (Hedges' g = −0.93 & −0.8) and verbal reasoning (g = −0.74 & −0.68) [48,50]. Cerebellar tumors in particular present with a range of speech and language deficits, including dysfluencies, slowed speech and reduced verbal abilities [81,82], associated with posterior fossa syndrome as discussed above.

PBT survivors are also at risk for sensory and motor impairments that can have a negative downstream impact on learning, academics, communication and social success. For instance, survivors may display early or delayed onset hearing loss attributed to the ototoxic effects of specific chemotherapy agents, as well as potential radiation-related damage to auditory structures [24,25,83].

3.2. Psychosocial Outcomes

Even though psychological problems after PBT have received less research attention than neurocognitive dysfunction, evidence suggests survivors are at greater risk for depression, anxiety, suicidal ideation and behavior problems relative to the general population [84]. Sense of well-being [85], family functioning [86], parent-child health related communication [87] and social involvement [88] have all been implicated as areas of risk after PBT.

Psychological problems and their prevalence rates are highly variable across samples, which impedes conclusive statements regarding patterns of psychological outcome [89]. Social deficits are well established in this population, including low social competence relative to typically developing children, siblings and survivors of non-CNS cancers [90,91]. Causation is unclear, although potential contributors include level of social skill development, functional or sensorimotor deficits, separation from peers and social networks, temperament and specific neurocognitive deficits such as decreased cognitive ability and attention [92,93].

The examination of individual differences and their impact on PBT survivors' psychological health has received some attention. As an example, neurocognitive dysfunction has been consistently associated with emotional and behavioral health [94–96]. Particular aspects of neurocognitive dysfunction, including executive function problems, present exponentially greater risk for emotional and behavioral health concerns in PBT survivors [97].

A recent conceptualization of the impact of childhood cancer on neurodevelopmental trajectory posits that the experience of childhood cancer is an early threat exposure that impacts psychological functioning and neural development [98] which helps unify the importance of addressing both psychosocial and neurocognitive late effects of PBT.

4. Interventions to Support PBT Survivors

Academic accommodations and modifications remain the primary educational support for academic difficulties resulting from neurocognitive late effects of tumor and treatment [99]. Indeed, special education utilization rates are especially high for this population relative to other types of childhood cancer [100]. Ongoing surveillance for neurocognitive and academic difficulties is considered standard of care and helps to inform academic supports and educational planning [41,42]. Career and vocational counseling may be helpful for PBT survivors who often face difficulties obtaining and maintaining employment when impacted by neurocognitive late effects [101,102].

In addition to educational supports, a number of cognitive training paradigms have targeted aspects of cognitive performance and academic achievement by attempting to enhance commonly affected functions including attention, working memory and processing speed [103]. Among the few studies that exist, results have largely been equivocal in terms of positive impact on brain tumor survivors' academic performance and outcome [42,103].

One PBT targeted paradigm utilized drill-oriented practice, metacognitive and learning skills acquisition and cognitive behavioral therapy [104,105] focused on improving attention and academic achievement. Statistically significant improvements were observed in a number of clinical measures. However, the relevance of these clinical test and rating improvements to the children's everyday performance was not established and this is unfortunately a common theme in research intervention programs targeting PBT late effects [103].

Recent attention has been devoted to computer-based training with the Cogmed [106] program to improve working memory using computer exercises along with regular coaching and support. Studies have suggested the intervention is feasible and acceptable in pediatric cancer survivors [107]. Randomized trials have shown performance improvements on clinical testing [108] and such improvements may be durable for months after the intervention [109]. However, studies supporting this program have not demonstrated specificity of computer-based training as the specific agent of improvement while controlling for level of support and coaching offered to participants [110]. More compelling evidence is needed before this intervention should be broadly recommended as efficacious in the PBT survivor population [103].

In addition to efforts focused on remediating specific neurocognitive deficits, other methods have addressed improving outcomes more globally in PBT survivors. For example, researchers have focused on indirect and contextual methods rooted in the premise that improving controllable external variables may hold promise for optimizing performance of brain-based functions that otherwise may not be amenable to direct intervention. For instance, training parents in behavioral modification, cognitive instructional methods and compensatory strategies to allow for ongoing intervention in the child's natural environment showed some efficacy in improving academic outcomes and warrants further attention [111]. As another example, a recent randomized study isolated improvements in situational motivation as associated with improved academic performance [112]. Situational and intrapersonal factors such as level of intrinsic achievement motivation and responsivity to external incentive may have a role in improving academic performance in PBT survivors. Finally, targeting health-related behaviors such as exercise, has shown promise in neural recovery and neurocognitive improvement and deserves further attention [113].

Pharmacological interventions to address neurocognitive late effects have been used with PBT survivors. Stimulant medications have been shown to improve aspects of attention in survivors but not intellectual functioning or academic skills [114]. In a small pilot study, donepezil—an acetylcholinesterase inhibitor—was shown to be feasible and to improve executive functioning and memory in childhood brain tumor survivors [115], justifying a more rigorous placebo controlled randomized trial. Modafinil has been examined as a possible medication to improve fatigue, cognitive functioning and mood in adult patients with primary brain tumors but its benefits did not exceed that of the placebo control group [116]. Pharmacologic prophylaxis to diminish neurotoxicity and preserve neurocognitive function after PBT treatment has shown preliminary utility in adults undergoing whole brain radiation [117] but no such prophylactic treatments have yet been systematically studied in children.

Finally, psychological interventions have demonstrated efficacy in ameliorating PBT survivors' behavioral and emotional health problems [118,119]. However, psychological referral standards have yet to be established and there is a clear disconnect in that the number of reported concerns far exceeds the frequency of referral for psychological services [96,120].

Figure 2 provides an overview of the recommended clinical management of neuropsychological late effects of PBT survivors, inclusive of ongoing clinical surveillance through individualized treatment planning. Because of the heterogeneity of factors affecting outcomes and the diversity of outcomes themselves, there currently is not a singular consensus pathway, timeline, or group of identified supports recommended for all patients. Ongoing surveillance for neurocognitive late effects is essential to engaging subsequent clinical neuropsychological assessment and treatment planning to optimize outcomes for PBT survivors.

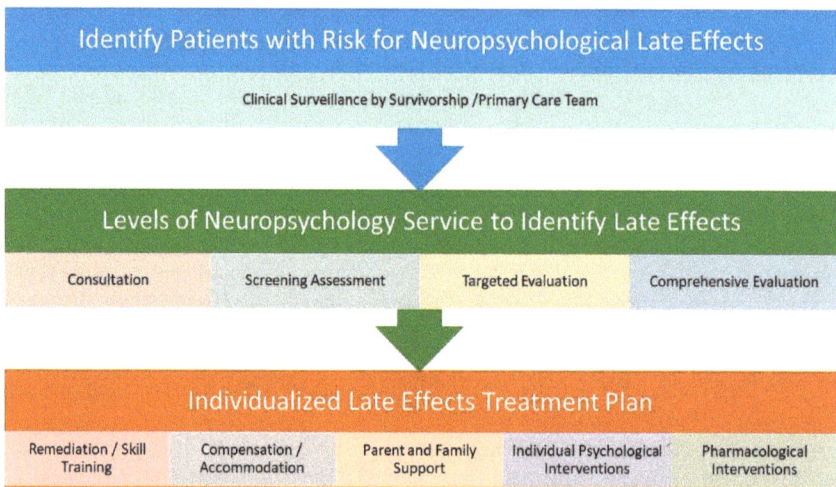

Figure 2. Clinical management of neuropsychological late effects after PBT.

5. Conclusions

PBT outcomes research is challenging for numerous reasons and it is within this context that the bulk of neurocognitive and psychosocial outcomes research should be understood. Small base rates of specific pediatric tumor types have often resulted in small research samples. A common amelioration of this challenge has been to include mixed types of pediatric cancers and treatment paradigms, though this then contributes to variable rates of reporting of things like psychological difficulties specific to tumor variables and treatment patterns [121]. Accruing patients over long periods of time is another potential remedy, though changes in treatment-related variables (e.g., changes in chemotherapy and/or radiation protocols) may result in incomparable samples over time. Non-medical and demographic factors known to correlate with cognitive and psychosocial functioning (e.g., family functioning, socioeconomic factors) are often unaccounted for and the unique nature of PBT research complicates identification of an appropriate "control" group. While the majority of PBT outcomes studies are cross-sectional, those that are longitudinal often suffer from a lack of clearly defined or valid baseline to which later results can be compared [122]. Finally, the needs of patient care are sometimes at odds with scientific rigor, as providing clinically useful information to families and providers may not align with consistency of data gathering.

While many challenges exist in studying this population, there have been improvements and innovations over the past several decades in terms of treatment paradigms to spare neurocognitive and psychosocial functioning in PBT survivors that have led to meaningful and even dramatic improvements in long-term outcome. Largely these improvements have resulted from refined treatment protocols that have reduced neurotoxicity of treatments for PBT and unfortunately less progress has been achieved in terms of intervention to improve late effects experienced by PBT survivors.

Nonetheless, as summarized above, clinicians and researchers alike should note the development of several promising domains that warrant more attention and provide potentially fruitful topics for future clinical research. While educational support through schools is considered standard of care, there is significant opportunity to improve educational programming and support to optimize academic outcomes. Further, novel intervention paradigms have shown some promise in recent years including direct training of neurocognitive functions affected by treatment for PBT. Individual factors such as intrinsic motivation and resilience are now being considered in terms of their relationship to neuropsychological late effects. Recent work has demonstrated potential efficacy of parent and family support as a way to ameliorate late effects. Pharmacological interventions have only recently been explored and clearly there are opportunities for collaborative clinical research to investigate efficacy of medications to improve neurocognitive function after treatment for PBT. Efforts investigating the impact of health-related behaviors such as nutrition and exercise on outcomes from PBT and its treatment are in their infancy and additional research in this area may help identify cost effective and readily accessible ways to improve neurocognitive functions in the PBT population. Finally, the role of preventative methods to reduce late effects burden is only now being explored and may represent a significant opportunity to improve outcomes for PBT survivors.

Optimal clinical management of neuropsychological late effects after treatment for PBT begins with awareness of the need to monitor this population at high risk for neuropsychological deficits. Unfortunately access to appropriate neuropsychological surveillance, evaluation and intervention remains inconsistent for PBT survivors [41,42]. Moreover, we are still in the beginning stages of determining effective strategies for implementing proposed standards of neuropsychological care, meeting patient need within the current healthcare climate and resource constraints common in various clinical settings. Advocacy for improved access to surveillance and care for neuropsychological late effects is the shared responsibility of researchers and clinicians working with this population to bring to the fore the needs of this vulnerable population as well as to establish efficacy of new and innovative interventions to improve outcomes.

Finally, in addition to efforts to improve surveillance and intervention paradigms and access to care, innovations in the basic conceptualization of mechanisms of cognitive impairment in pediatric cancer, such as examining structural connectome organization implicated in efficiency of information processing [123], may further refine our understanding and detection of late effects of PBT and its treatment. What is clear is that preventing, managing and remediating late neurocognitive and psychosocial late effects for PBT survivors is going to require innovation and problem-solving among numerous basic and applied scientific disciplines.

Author Contributions: Conceptualization, P.L.S., M.A.A., S.K.P., N.P.S., R.S.R.; Writing-Original Draft Preparation, P.L.S., M.A.A., S.K.P., N.P.S., R.S.R.; Writing-Review & Editing, P.L.S., M.A.A., S.K.P., N.P.S., R.S.R.

Funding: This review received no external funding.

Conflicts of Interest: The authors declare no conflict of interest.

References

1. Olsson, I.T.; Perrin, S.; Lundgren, J.; Hjorth, L.; Johanson, A. Long-term cognitive sequelae after pediatric brain tumor related to medical risk factors, age, and sex. *Pediatr. Neurol.* **2014**, *51*, 515–521. [CrossRef] [PubMed]

2. Palmer, S.L.; Armstrong, C.; Onar-Thomas, A.; Wu, S.; Wallace, D.; Bonner, M.J.; Schreiber, J.; Swain, M.; Chapieski, L.; Mabbott, D.; et al. Processing speed, attention, and working memory after treatment for medulloblastoma: An international, prospective, and longitudinal study. *J. Clin. Oncol.* **2013**, *31*, 3494–3500. [CrossRef] [PubMed]
3. Schreiber, J.E.; Gurney, J.G.; Palmer, S.L.; Bass, J.K.; Wang, M.; Chen, S.; Zhang, H.; Swain, M.; Chapieski, M.L.; Bonner, M.J. Examination of risk factors for intellectual and academic outcomes following treatment for pediatric medulloblastoma. *Neuro-Oncology* **2014**, *16*, 1129–1136. [CrossRef] [PubMed]
4. Di Rocco, F.; Jucá, C.E.; Zerah, M.; Sainte-Rose, C. Endoscopic third ventriculostomy and posterior fossa tumors. *World Neurosurg.* **2013**, *79*, S18.e15–S18.e19. [CrossRef] [PubMed]
5. Hardy, K.K.; Bonner, M.J.; Willard, V.W.; Watral, M.A.; Gururangan, S. Hydrocephalus as a possible additional contributor to cognitive outcome in survivors of pediatric medulloblastoma. *Psycho-Oncol. J. Psychol. Soc. Behav. Dimens. Cancer* **2008**, *17*, 1157–1161. [CrossRef] [PubMed]
6. Duffner, P.K. Risk factors for cognitive decline in children treated for brain tumors. *Eur. J. Paediatr. Neurol.* **2010**, *14*, 106–115. [CrossRef] [PubMed]
7. Patel, S.; Mullins, W.; O'neil, S.; Wilson, K. Neuropsychological differences between survivors of supratentorial and infratentorial brain tumours. *J. Intell. Disabil. Res.* **2011**, *55*, 30–40. [CrossRef] [PubMed]
8. Beebe, D.W.; Ris, M.D.; Armstrong, F.D.; Fontanesi, J.; Mulhern, R.; Holmes, E.; Wisoff, J.H. Cognitive and adaptive outcome in low-grade pediatric cerebellar astrocytomas: Evidence of diminished cognitive and adaptive functioning in National Collaborative Research Studies (CCG 9891/POG 9130). *J. Clin. Oncol.* **2005**, *23*, 5198–5204. [CrossRef] [PubMed]
9. Jalali, R.; Mallick, I.; Dutta, D.; Goswami, S.; Gupta, T.; Munshi, A.; Deshpande, D.; Sarin, R. Factors influencing neurocognitive outcomes in young patients with benign and low-grade brain tumors treated with stereotactic conformal radiotherapy. *Int. J. Radiat. Oncol. Biol. Phys.* **2010**, *77*, 974–979. [CrossRef] [PubMed]
10. Ris, M.D.; Beebe, D.W.; Armstrong, F.D.; Fontanesi, J.; Holmes, E.; Sanford, R.A.; Wisoff, J.H. Cognitive and adaptive outcome in extracerebellar low-grade brain tumors in children: A report from the Children's Oncology Group. *J. Clin. Oncol.* **2008**, *26*, 4765. [CrossRef] [PubMed]
11. Avula, S.; Spiteri, M.; Kumar, R.; Lewis, E.; Harave, S.; Windridge, D.; Ong, C.; Pizer, B. Post-operative pediatric cerebellar mutism syndrome and its association with hypertrophic olivary degeneration. *Quant. Imaging Med. Surg.* **2016**, *6*, 535. [CrossRef] [PubMed]
12. Schreiber, J.E.; Palmer, S.L.; Conklin, H.M.; Mabbott, D.J.; Swain, M.A.; Bonner, M.J.; Chapieski, M.L.; Huang, L.; Zhang, H.; Gajjar, A. Posterior fossa syndrome and long-term neuropsychological outcomes among children treated for medulloblastoma on a multi-institutional, prospective study. *Neuro-Oncology* **2017**, *19*, 1673–1682. [CrossRef] [PubMed]
13. De Ruiter, M.A.; van Mourik, R.; Schouten-van Meeteren, A.Y.; Grootenhuis, M.A.; Oosterlaan, J. Neurocognitive consequences of a paediatric brain tumour and its treatment: A meta-analysis. *Dev. Med. Child. Neurol.* **2013**, *55*, 408–417. [CrossRef] [PubMed]
14. Spiegler, B.J.; Bouffet, E.; Greenberg, M.L.; Rutka, J.T.; Mabbott, D.J. Change in neurocognitive functioning after treatment with cranial radiation in childhood. *J. Clin. Oncol.* **2004**, *22*, 706–713. [CrossRef] [PubMed]
15. Fouladi, M.; Chintagumpala, M.; Laningham, F.H.; Ashley, D.; Kellie, S.J.; Langston, J.W.; McCluggage, C.W.; Woo, S.; Kocak, M.; Krull, K. White matter lesions detected by magnetic resonance imaging after radiotherapy and high-dose chemotherapy in children with medulloblastoma or primitive neuroectodermal tumor. *J. Clin. Oncol.* **2004**, *22*, 4551–4560. [CrossRef] [PubMed]
16. Jacola, L.M.; Ashford, J.M.; Reddick, W.E.; Glass, J.O.; Ogg, R.J.; Merchant, T.E.; Conklin, H.M. The relationship between working memory and cerebral white matter volume in survivors of childhood brain tumors treated with conformal radiation therapy. *J. Neuro-Oncol.* **2014**, *119*, 197–205. [CrossRef] [PubMed]
17. Reddick, W.E.; Glass, J.O.; Palmer, S.L.; Wu, S.; Gajjar, A.; Langston, J.W.; Kun, L.E.; Xiong, X.; Mulhern, R.K. Atypical white matter volume development in children following craniospinal irradiation. *Neuro-Oncology* **2005**, *7*, 12–19. [CrossRef] [PubMed]
18. Monje, M.L.; Vogel, H.; Masek, M.; Ligon, K.L.; Fisher, P.G.; Palmer, T.D. Impaired human hippocampal neurogenesis after treatment for central nervous system malignancies. *Ann. Neurol. Off. J. Am. Neurol. Assoc. Child Neurol. Soc.* **2007**, *62*, 515–520. [CrossRef] [PubMed]

19. Nagel, B.J.; Delis, D.C.; Palmer, S.L.; Reeves, C.; Gajjar, A.; Mulhern, R.K. Early patterns of verbal memory impairment in children treated for medulloblastoma. *Neuropsychology* **2006**, *20*, 105. [CrossRef] [PubMed]

20. Law, N.; Bouffet, E.; Laughlin, S.; Laperriere, N.; Brière, M.-E.; Strother, D.; McConnell, D.; Hukin, J.; Fryer, C.; Rockel, C. Cerebello–thalamo–cerebral connections in pediatric brain tumor patients: Impact on working memory. *Neuroimage* **2011**, *56*, 2238–2248. [CrossRef] [PubMed]

21. Taylor, O.A.; Hockenberry, M.J.; McCarthy, K.; Gundy, P.; Montgomery, D.; Ross, A.; Scheurer, M.E.; Moore, I.M. Evaluation of biomarkers of oxidative stress and apoptosis in patients with severe methotrexate neurotoxicity: A case series. *J. Pediatr. Oncol. Nurs.* **2015**, *32*, 320–325. [CrossRef] [PubMed]

22. Rutkowski, S.; Bode, U.; Deinlein, F.; Ottensmeier, H.; Warmuth-Metz, M.; Soerensen, N.; Graf, N.; Emser, A.; Pietsch, T.; Wolff, J.E. Treatment of early childhood medulloblastoma by postoperative chemotherapy alone. *N. Engl. J. Med.* **2005**, *352*, 978–986. [CrossRef] [PubMed]

23. Verstappen, C.C.; Heimans, J.J.; Hoekman, K.; Postma, T.J. Neurotoxic complications of chemotherapy in patients with cancer. *Drugs* **2003**, *63*, 1549–1563. [CrossRef] [PubMed]

24. McHaney, V.A.; Thibadoux, G.; Hayes, F.A.; Green, A.A. Hearing loss in children receiving cisplatin chemotherapy. *J. Pediatr.* **1983**, *102*, 314–317. [CrossRef]

25. Warrier, R.; Chauhan, A.; Davluri, M.; Tedesco, S.L.; Nadell, J.; Craver, R. Cisplatin and cranial irradiation-related hearing loss in children. *Ochsner J.* **2012**, *12*, 191–196. [PubMed]

26. Bull, K.S.; Spoudeas, H.A.; Yadegarfar, G.; Kennedy, C.R. Reduction of health status 7 years after addition of chemotherapy to craniospinal irradiation for medulloblastoma: A follow-up study in PNET 3 trial survivors—On behalf of the CCLG (formerly UKCCSG). *J. Clin. Oncol.* **2007**, *25*, 4239–4245. [CrossRef] [PubMed]

27. Di Pinto, M.; Conklin, H.M.; Li, C.; Merchant, T.E. Learning and memory following conformal radiation therapy for pediatric craniopharyngioma and low-grade glioma. *Int. J. Radiat. Oncol. Biol. Phys.* **2012**, *84*, e363–e369. [CrossRef] [PubMed]

28. Ris, M.D.; Packer, R.; Goldwein, J.; Jones-Wallace, D.; Boyett, J.M. Intellectual outcome after reduced-dose radiation therapy plus adjuvant chemotherapy for medulloblastoma: A Children's Cancer Group study. *J. Clin. Oncol.* **2001**, *19*, 3470–3476. [CrossRef] [PubMed]

29. Andreotti, C.; Root, J.C.; Ahles, T.A.; McEwen, B.S.; Compas, B.E. Cancer, coping, and cognition: A model for the role of stress reactivity in cancer-related cognitive decline. *Psycho-Oncology* **2015**, *24*, 617–623. [CrossRef] [PubMed]

30. Jain, N.; Brouwers, P.; Okcu, M.F.; Cirino, P.T.; Krull, K.R. Sex-specific attention problems in long-term survivors of pediatric acute lymphoblastic leukemia. *Cancer Interdiscip. Int. J. Am. Cancer Soc.* **2009**, *115*, 4238–4245. [CrossRef] [PubMed]

31. Palmer, S.L. Neurodevelopmental impact on children treated for medulloblastoma: A review and proposed conceptual model. *Dev. Disabil. Res. Rev.* **2008**, *14*, 203–210. [CrossRef] [PubMed]

32. Patel, S.K.; Fernandez, N.; Dekel, N.; Turk, A.; Meier, A.; Ross, P.; Rosenthal, J. Socioeconomic status as a possible moderator of neurocognitive outcomes in children with cancer. *Psycho-Oncology* **2016**, *25*, 115–118. [CrossRef] [PubMed]

33. Robinson, K.E.; Wolfe, K.R.; Yeates, K.O.; Mahone, E.M.; Cecil, K.M.; Ris, M.D. Predictors of adaptive functioning and psychosocial adjustment in children with pediatric brain tumor: A report from the brain radiation investigative study consortium. *Pediatr. Blood Cancer* **2015**, *62*, 509–516. [CrossRef] [PubMed]

34. French, A.E.; Tsangaris, E.; Barrera, M.; Guger, S.; Brown, R.; Urbach, S.; Stephens, D.; Nathan, P.C. School attendance in childhood cancer survivors and their siblings. *J. Pediatr.* **2013**, *162*, 160–165. [CrossRef] [PubMed]

35. Butler, R.W.; Haser, J.K. Neurocognitive effects of treatment for childhood cancer. *Ment. Retard. Dev. Disabil. Res. Rev.* **2006**, *12*, 184–191. [CrossRef] [PubMed]

36. Palmer, S.L.; Reddick, W.E.; Gajjar, A. Understanding the cognitive impact on children who are treated for medulloblastoma. *J. Pediatr. Psychol.* **2007**, *32*, 1040–1049. [CrossRef] [PubMed]

37. Mabbott, D.J.; Spiegler, B.J.; Greenberg, M.L.; Rutka, J.T.; Hyder, D.J.; Bouffet, E. Serial evaluation of academic and behavioral outcome after treatment with cranial radiation in childhood. *J. Clin. Oncol.* **2005**, *23*, 2256–2263. [CrossRef] [PubMed]

38. Mabbott, D.J.; Penkman, L.; Witol, A.; Strother, D.; Bouffet, E. Core neurocognitive functions in children treated for posterior fossa tumors. *Neuropsychology* **2008**, *22*, 159. [CrossRef] [PubMed]

39. Dennis, M. Childhood medical disorders and cognitive impairment: Biological risk, time, development, and reserve. In *Pediatric Neuropsychology: Research, Theory, and Practice*; Yeates, K.O., Ris, M.D., Taylor, H.G., Eds.; Guilford Press: New York, NY, USA, 2000; pp. 3–22.

40. Baum, K.T.; Powell, S.K.; Jacobson, L.A.; Gragert, M.N.; Janzen, L.A.; Paltin, I.; Rey-Casserly, C.M.; Wilkening, G.N. Implementing guidelines: Proposed definitions of neuropsychology services in pediatric oncology. *Pediatr. Blood Cancer* **2017**, *64*, e26446. [CrossRef] [PubMed]

41. Nathan, P.C.; Patel, S.K.; Dilley, K.; Goldsby, R.; Harvey, J.; Jacobsen, C.; Kadan-Lottick, N.; McKinley, K.; Millham, A.K.; Moore, I.; et al. Guidelines for identification of, advocacy for, and intervention in neurocognitive problems in survivors of childhood cancer: A report from the Children's Oncology Group. *Arch. Pediatr. Adolesc. Med.* **2007**, *161*, 798–806. [CrossRef] [PubMed]

42. Annett, R.D.; Patel, S.K.; Phipps, S. Monitoring and Assessment of Neuropsychological Outcomes as a Standard of Care in Pediatric Oncology. *Pediatr. Blood Cancer* **2015**, *62* (Suppl. 5), S460–S513. [CrossRef] [PubMed]

43. Bledsoe, J.C. Effects of Cranial Radiation on Structural and Functional Brain Development in Pediatric Brain Tumors. *J. Pediatr. Neuropsychol.* **2016**, *2*, 3–13. [CrossRef]

44. Grill, J.; Renaux, V.K.; Bulteau, C.; Viguier, D.; Levy-Piebois, C.; Sainte-Rose, C.; Dellatolas, G.; Raquin, M.A.; Jambaque, I.; Kalifa, C. Long-term intellectual outcome in children with posterior fossa tumors according to radiation doses and volumes. *Int. J. Radiat. Oncol. Biol. Phys.* **1999**, *45*, 137–145. [CrossRef]

45. Moxon-Emre, I.; Bouffet, E.; Taylor, M.D.; Laperriere, N.; Scantlebury, N.; Law, N.; Spiegler, B.J.; Malkin, D.; Janzen, L.; Mabbott, D. Impact of craniospinal dose, boost volume, and neurologic complications on intellectual outcome in patients with medulloblastoma. *J. Clin. Oncol.* **2014**, *32*, 1760–1768. [CrossRef] [PubMed]

46. Chevignard, M.; Câmara-Costa, H.; Doz, F.; Dellatolas, G. Core deficits and quality of survival after childhood medulloblastoma: A review. *Neuro-Oncol. Pract.* **2017**, *4*, 82–97. [CrossRef]

47. Brinkman, T.M.; Reddick, W.E.; Luxton, J.; Glass, J.O.; Sabin, N.D.; Srivastava, D.K.; Robison, L.L.; Hudson, M.M.; Krull, K.R. Cerebral white matter integrity and executive function in adult survivors of childhood medulloblastoma. *Neuro-Oncology* **2012**, *14* (Suppl. 4), iv25–iv36. [CrossRef] [PubMed]

48. Robinson, K.E.; Kuttesch, J.F.; Champion, J.E.; Andreotti, C.F.; Hipp, D.W.; Bettis, A.; Barnwell, A.; Compas, B.E. A quantitative meta-analysis of neurocognitive sequelae in survivors of pediatric brain tumors. *Pediatr. Blood Cancer* **2010**, *55*, 525–531. [CrossRef] [PubMed]

49. King, T.Z.; Ailion, A.S.; Fox, M.E.; Hufstetler, S.M. Neurodevelopmental model of long-term outcomes of adult survivors of childhood brain tumors. *Child Neuropsychol.* **2017**, 1–21. [CrossRef] [PubMed]

50. Robinson, K.E.; Fraley, C.E.; Pearson, M.M.; Kuttesch, J.F., Jr.; Compas, B.E. Neurocognitive late effects of pediatric brain tumors of the posterior fossa: A quantitative review. *J. Int. Neuropsychol. Soc.* **2013**, *19*, 44–53. [CrossRef] [PubMed]

51. Palmer, S.L.; Goloubeva, O.; Reddick, W.E.; Glass, J.O.; Gajjar, A.; Kun, L.; Merchant, T.E.; Mulhern, R.K. Patterns of intellectual development among survivors of pediatric medulloblastoma: A longitudinal analysis. *J. Clin. Oncol.* **2001**, *19*, 2302–2308. [CrossRef] [PubMed]

52. Palmer, S.L.; Gajjar, A.; Reddick, W.E.; Glass, J.O.; Kun, L.E.; Wu, S.; Xiong, X.; Mulhern, R.K. Predicting intellectual outcome among children treated with 35–40 Gy craniospinal irradiation for medulloblastoma. *Neuropsychology* **2003**, *17*, 548–555. [CrossRef] [PubMed]

53. Radcliffe, J.; Bunin, G.R.; Sutton, L.N.; Goldwein, J.W.; Phillips, P.C. Cognitive deficits in long-term survivors of childhood medulloblastoma and other noncortical tumors: Age-dependent effects of whole brain radiation. *Int. J. Dev. Neurosci.* **1994**, *12*, 327–334. [CrossRef]

54. Kieffer-Renaux, V.; Bulteau, C.; Grill, J.; Kalifa, C.; Viguier, D.; Jambaque, I. Patterns of neuropsychological deficits in children with medulloblastoma according to craniospatial irradiation doses. *Dev. Med. Child Neurol.* **2000**, *42*, 741–745. [CrossRef] [PubMed]

55. Mulhern, R.K.; Kepner, J.L.; Thomas, P.R.; Armstrong, F.D.; Friedman, H.S.; Kun, L.E. Neuropsychologic functioning of survivors of childhood medulloblastoma randomized to receive conventional or reduced-dose craniospinal irradiation: A Pediatric Oncology Group study. *J. Clin. Oncol.* **1998**, *16*, 1723–1728. [CrossRef] [PubMed]

56. Wilson, V.C.; McDonough, J.; Tochner, Z. Proton beam irradiation in pediatric oncology: An overview. *J. Pediatr. Hematol. Oncol.* **2005**, *27*, 444–448. [CrossRef] [PubMed]

57. Warren, E.; Child, A.; Cirino, P.; Grosshans, D.; Mahajan, A.; Paulino, A.; Chintagumpala, M.; Okcu, F.; Douglas Ris, M.; Minard, C. QOL-42. Better social, cognitive, and academic outcomes among pediatric brain tumor survivors treated with proton versus photon radiation therapy. *Neuro-Oncology* **2018**, *20*, i166. [CrossRef]

58. Kahalley, L.S.; Ris, M.D.; Grosshans, D.R.; Okcu, M.F.; Paulino, A.C.; Chintagumpala, M.; Moore, B.D.; Guffey, D.; Minard, C.G.; Stancel, H.H.; et al. Comparing Intelligence Quotient Change After Treatment With Proton Versus Photon Radiation Therapy for Pediatric Brain Tumors. *J. Clin. Oncol.* **2016**, *34*, 1043–1049. [CrossRef] [PubMed]

59. Antonini, T.N.; Ris, M.D.; Grosshans, D.R.; Mahajan, A.; Okcu, M.F.; Chintagumpala, M.; Paulino, A.; Child, A.E.; Orobio, J.; Stancel, H.H.; et al. Attention, processing speed, and executive functioning in pediatric brain tumor survivors treated with proton beam radiation therapy. *Radiother. Oncol.* **2017**, *124*, 89–97. [CrossRef] [PubMed]

60. Conklin, H.M.; Ashford, J.M.; Howarth, R.A.; Merchant, T.E.; Ogg, R.J.; Santana, V.M.; Reddick, W.E.; Wu, S.; Xiong, X. Working memory performance among childhood brain tumor survivors. *J. Int. Neuropsychol. Soc.* **2012**, *18*, 996–1005. [CrossRef] [PubMed]

61. Reddick, W.E.; White, H.A.; Glass, J.O.; Wheeler, G.C.; Thompson, S.J.; Gajjar, A.; Leigh, L.; Mulhern, R.K. Developmental model relating white matter volume to neurocognitive deficits in pediatric brain tumor survivors. *Cancer* **2003**, *97*, 2512–2519. [CrossRef] [PubMed]

62. Kahalley, L.S.; Conklin, H.M.; Tyc, V.L.; Hudson, M.M.; Wilson, S.J.; Wu, S.; Xiong, X.; Hinds, P.S. Slower processing speed after treatment for pediatric brain tumor and acute lymphoblastic leukemia. *Psychooncology* **2013**, *22*, 1979–1986. [CrossRef] [PubMed]

63. Schatz, J.; Kramer, J.H.; Ablin, A.; Matthay, K.K. Processing speed, working memory, and IQ: A developmental model of cognitive deficits following cranial radiation therapy. *Neuropsychology* **2000**, *14*, 189–200. [CrossRef] [PubMed]

64. Fry, A.F.; Hale, S. Relationships among processing speed, working memory, and fluid intelligence in children. *Biol. Psychol.* **2000**, *54*, 1–34. [CrossRef]

65. Holland, A.A.; Hughes, C.W.; Stavinoha, P.L. School competence and fluent academic performance: Informing assessment of educational outcomes in survivors of pediatric medulloblastoma. *Appl. Neuropsychol. Child* **2015**, *4*, 249–256. [CrossRef] [PubMed]

66. Roddy, E.; Mueller, S. Late effects of treatment of pediatric central nervous system tumors. *J. Child Neurol.* **2016**, *31*, 237–254. [CrossRef] [PubMed]

67. Starowicz-Filip, A.; Chrobak, A.A.; Milczarek, O.; Kwiatkowski, S. The visuospatial functions in children after cerebellar low-grade astrocytoma surgery: A contribution to the pediatric neuropsychology of the cerebellum. *J. Neuropsychol.* **2017**, *11*, 201–221. [CrossRef] [PubMed]

68. Koustenis, E.; Driever, P.H.; de Sonneville, L.; Rueckriegel, S.M. Executive function deficits in pediatric cerebellar tumor survivors. *Eur. J. Paediatr. Neurol.* **2016**, *20*, 25–37. [CrossRef] [PubMed]

69. Robison, L.L.; Hudson, M.M. Survivors of childhood and adolescent cancer: Life-long risks and responsibilities. *Nat. Rev. Cancer* **2014**, *14*, 61. [CrossRef] [PubMed]

70. Wochos, G.; Semerjian, C.; Walsh, K.S. Differences in parent and teacher rating of everyday executive function in pediatric brain tumor survivors. *Clin. Neuropsychol.* **2014**, *28*, 1243–1257. [CrossRef] [PubMed]

71. Wolfe, K.R.; Walsh, K.S.; Reynolds, N.C.; Mitchell, F.; Reddy, A.T.; Paltin, I.; Madan-Swain, A. Executive functions and social skills in survivors of pediatric brain tumor. *Child Neuropsychol.* **2013**, *19*, 370–384. [CrossRef] [PubMed]

72. Armstrong, G.T.; Liu, Q.; Yasui, Y.; Huang, S.; Ness, K.K.; Leisenring, W.; Hudson, M.M.; Donaldson, S.S.; King, A.A.; Stovall, M.; et al. Long-term outcomes among adult survivors of childhood central nervous system malignancies in the Childhood Cancer Survivor Study. *J. Natl. Cancer Inst.* **2009**, *101*, 946–958. [CrossRef] [PubMed]

73. Ribi, K.; Relly, C.; Landolt, M.A.; Alber, F.D.; Boltshauser, E.; Grotzer, M.A. Outcome of medulloblastoma in children: Long-term complications and quality of life. *Neuropediatrics* **2005**, *36*, 357–365. [CrossRef] [PubMed]

74. Ronning, C.; Sundet, K.; Due-Tonnessen, B.; Lundar, T.; Helseth, E. Persistent cognitive dysfunction secondary to cerebellar injury in patients treated for posterior fossa tumors in childhood. *Pediatr. Neurosurg.* **2005**, *41*, 15–21. [CrossRef] [PubMed]

75. Roncadin, C.; Dennis, M.; Greenberg, M.L.; Spiegler, B.J. Adverse medical events associated with childhood cerebellar astrocytomas and medulloblastomas: Natural history and relation to very long-term neurobehavioral outcome. *Childs Nerv. Syst.* **2008**, *24*, 995–1002. [CrossRef] [PubMed]
76. Maddrey, A.M.; Bergeron, J.A.; Lombardo, E.R.; McDonald, N.K.; Mulne, A.F.; Barenberg, P.D.; Bowers, D.C. Neuropsychological performance and quality of life of 10 year survivors of childhood medulloblastoma. *J. Neurooncol.* **2005**, *72*, 245–253. [CrossRef] [PubMed]
77. Edelstein, K.; Spiegler, B.J.; Fung, S.; Panzarella, T.; Mabbott, D.J.; Jewitt, N.; D'Agostino, N.M.; Mason, W.P.; Bouffet, E.; Tabori, U.; et al. Early aging in adult survivors of childhood medulloblastoma: Long-term neurocognitive, functional, and physical outcomes. *Neuro-Oncology* **2011**, *13*, 536–545. [CrossRef] [PubMed]
78. Mulhern, R.K.; Palmer, S.L.; Reddick, W.E.; Glass, J.O.; Kun, L.E.; Taylor, J.; Langston, J.; Gajjar, A. Risks of young age for selected neurocognitive deficits in medulloblastoma are associated with white matter loss. *J. Clin. Oncol.* **2001**, *19*, 472–479. [CrossRef] [PubMed]
79. Hocking, M.C.; Hobbie, W.L.; Deatrick, J.A.; Hardie, T.L.; Barakat, L.P. Family functioning mediates the association between neurocognitive functioning and health-related quality of life in young adult survivors of childhood brain tumors. *J. Adolesc. Young Adult Oncol.* **2015**, *4*, 18–25. [CrossRef] [PubMed]
80. Redmond, K.J.; Mahone, E.M.; Terezakis, S.; Ishaq, O.; Ford, E.; McNutt, T.; Kleinberg, L.; Cohen, K.J.; Wharam, M.; Horska, A. Association between radiation dose to neuronal progenitor cell niches and temporal lobes and performance on neuropsychological testing in children: A prospective study. *Neuro-Oncology* **2013**, *15*, 360–369. [CrossRef] [PubMed]
81. Huber, J.F.; Bradley, K.; Spiegler, B.; Dennis, M. Long-term neuromotor speech deficits in survivors of childhood posterior fossa tumors: Effects of tumor type, radiation, age at diagnosis, and survival years. *J. Child Neurol.* **2007**, *22*, 848–854. [CrossRef] [PubMed]
82. Catsman-Berrevoets, C.E.; Aarsen, F.K. The spectrum of neurobehavioural deficits in the Posterior Fossa Syndrome in children after cerebellar tumour surgery. *Cortex* **2010**, *46*, 933–946. [CrossRef] [PubMed]
83. Hua, C.; Bass, J.K.; Khan, R.; Kun, L.E.; Merchant, T.E. Hearing loss after radiotherapy for pediatric brain tumors: Effect of cochlear dose. *Int. J. Radiat. Oncol. Biol. Phys.* **2008**, *72*, 892–899. [CrossRef] [PubMed]
84. Shah, S.S.; Dellarole, A.; Peterson, E.C.; Bregy, A.; Komotar, R.; Harvey, P.D.; Elhammady, M.S. Long-term psychiatric outcomes in pediatric brain tumor survivors. *Child's Nerv. Syst.* **2015**, *31*, 653–663. [CrossRef] [PubMed]
85. Memmesheimer, R.M.; Lange, K.; Dölle, M.; Heger, S.; Mueller, I. Psychological well-being and independent living of young adults with childhood-onset craniopharyngioma. *Dev. Med. Child Neurol.* **2017**, *59*, 829–836. [CrossRef] [PubMed]
86. Quast, L.F.; Phillips, P.C.; Li, Y.; Kazak, A.E.; Barakat, L.P.; Hocking, M.C. A prospective study of family predictors of health-related quality of life in pediatric brain tumor survivors. *Pediatr. Blood Cancer* **2018**, *65*, e26976. [CrossRef] [PubMed]
87. Adduci, A.; Jankovic, M.; Strazzer, S.; Massimino, M.; Clerici, C.; Poggi, G. Parent–child communication and psychological adjustment in children with a brain tumor. *Pediatr. Blood Cancer* **2012**, *59*, 290–294. [CrossRef] [PubMed]
88. De Ruiter, M.A.; Schouten-van Meeteren, A.Y.N.; van Vuurden, D.G.; Maurice-Stam, H.; Gidding, C.; Beek, L.R.; Granzen, B.; Oosterlaan, J.; Grootenhuis, M.A. Psychosocial profile of pediatric brain tumor survivors with neurocognitive complaints. *Qual. Life Res.* **2016**, *25*, 435–446. [CrossRef] [PubMed]
89. Fuemmeler, B.F.; Elkin, T.D.; Mullins, L.L. Survivors of childhood brain tumors: Behavioral, emotional, and social adjustment. *Clin. Psychol. Rev.* **2002**, *22*, 547–585. [CrossRef]
90. Emond, A.; Edwards, L.; Peacock, S.; Norman, C.; Evangeli, M. Social competence in children and young people treated for a brain tumour. *Support. Care Cancer* **2016**, *24*, 4587–4595. [CrossRef] [PubMed]
91. Schulte, F. Social competence in pediatric brain tumor survivors: Breadth versus depth. *Curr. Opin. Oncol.* **2015**, *27*, 306–310. [CrossRef] [PubMed]
92. Holland, A.A.; Colaluca, B.; Bailey, L.; Stavinoha, P.L. Impact of attention on social functioning in pediatric medulloblastoma survivors. *Pediatr. Hematol. Oncol.* **2018**, *35*, 76–89. [CrossRef] [PubMed]
93. Sands, S.A.; Pasichow, K.P. Psychological and social impact of being a pediatric brain tumor survivor. In *Late Effects of Treatment for Brain Tumors*; Springer: New York, NY, USA, 2009; pp. 297–307.
94. Nassau, J.H.; Drotar, D. Social competence among children with central nervous system-related chronic health conditions: A review. *J. Pediatr. Psychol.* **1997**, *22*, 771–793. [CrossRef] [PubMed]

95. Poggi, G.; Liscio, M.; Galbiati, S.; Adduci, A.; Massimino, M.; Gandola, L.; Spreafico, F.; Clerici, C.A.; Fossati-Bellani, F.; Sommovigo, M. Brain tumors in children and adolescents: Cognitive and psychological disorders at different ages. *Psycho-Oncol. J. Psychol. Soc. Behav. Dimens. Cancer* **2005**, *14*, 386–395. [CrossRef] [PubMed]

96. Kahalley, L.S.; Wilson, S.J.; Tyc, V.L.; Conklin, H.M.; Hudson, M.M.; Wu, S.; Xiong, X.; Stancel, H.H.; Hinds, P.S. Are the psychological needs of adolescent survivors of pediatric cancer adequately identified and treated? *Psycho-Oncology* **2013**, *22*, 447–458. [CrossRef] [PubMed]

97. Ventura, L.M.; Grieco, J.A.; Evans, C.L.; Kuhlthau, K.A.; MacDonald, S.M.; Tarbell, N.J.; Yock, T.I.; Pulsifer, M.B. Executive functioning, academic skills, and quality of life in pediatric patients with brain tumors post-proton radiation therapy. *J. Neuro-Oncol.* **2018**, *137*, 119–126. [CrossRef] [PubMed]

98. Marusak, H.A.; Iadipaolo, A.S.; Harper, F.W.; Elrahal, F.; Taub, J.W.; Goldberg, E.; Rabinak, C.A. Neurodevelopmental consequences of pediatric cancer and its treatment: Applying an early adversity framework to understanding cognitive, behavioral, and emotional outcomes. *Neuropsychol. Rev.* **2017**, *28*, 123–175. [CrossRef] [PubMed]

99. Hay, G.H.; Nabors, M.; Sullivan, A.; Zygmund, A. Students with Pediatric Cancer: A Prescription for School Success. *Phys. Disabil. Educ. Relat. Serv.* **2015**, *34*, 1–13. [CrossRef]

100. Mitby, P.A.; Robison, L.L.; Whitton, J.A.; Zevon, M.A.; Gibbs, I.C.; Tersak, J.M.; Meadows, A.T.; Stovall, M.; Zeltzer, L.K.; Mertens, A.C.; et al. Utilization of special education services and educational attainment among long-term survivors of childhood cancer: A report from the Childhood Cancer Survivor Study. *Cancer* **2003**, *97*, 1115–1126. [CrossRef] [PubMed]

101. Kirchhoff, A.C.; Leisenring, W.; Krull, K.R.; Ness, K.K.; Friedman, D.L.; Armstrong, G.T.; Stovall, M.; Park, E.R.; Oeffinger, K.C.; Hudson, M.M.; et al. Unemployment among adult survivors of childhood cancer: A report from the childhood cancer survivor study. *Med. Care* **2010**, *48*, 1015–1025. [CrossRef] [PubMed]

102. Thompson, A.L.; Christiansen, H.L.; Elam, M.; Hoag, J.; Irwin, M.K.; Pao, M.; Voll, M.; Noll, R.B.; Kelly, K.P. Academic Continuity and School Reentry Support as a Standard of Care in Pediatric Oncology. *Pediatr. Blood Cancer* **2015**, *62* (Suppl. 5), S805–S817. [CrossRef] [PubMed]

103. Olson, K.; Sands, S.A. Cognitive training programs for childhood cancer patients and survivors: A critical review and future directions. *Child Neuropsychol.* **2016**, *22*, 509–536. [CrossRef] [PubMed]

104. Butler, R.W.; Copeland, D.R. Attentional processes and their remediation in children treated for cancer: A literature review and the development of a therapeutic approach. *J. Int. Neuropsychol. Soc.* **2002**, *8*, 115–124. [CrossRef] [PubMed]

105. Butler, R.W.; Copeland, D.R.; Fairclough, D.L.; Mulhern, R.K.; Katz, E.R.; Kazak, A.E.; Noll, R.B.; Patel, S.K.; Sahler, O.J.Z. A multicenter, randomized clinical trial of a cognitive remediation program for childhood survivors of a pediatric malignancy. *J. Consult. Clin. Psychol.* **2008**, *76*, 367. [CrossRef] [PubMed]

106. Pearson Education, Incorporated Cogmed. Available online: https://www.cogmed.com/ (accessed on 20 July 2018).

107. Hardy, K.K.; Willard, V.W.; Allen, T.M.; Bonner, M.J. Working memory training in survivors of pediatric cancer: A randomized pilot study. *Psycho-Oncology* **2013**, *22*, 1856–1865. [CrossRef] [PubMed]

108. Conklin, H.M.; Ogg, R.J.; Ashford, J.M.; Scoggins, M.A.; Zou, P.; Clark, K.N.; Martin-Elbahesh, K.; Hardy, K.K.; Merchant, T.E.; Jeha, S. Computerized cognitive training for amelioration of cognitive late effects among childhood cancer survivors: A randomized controlled trial. *J. Clin. Oncol.* **2015**, *33*, 3894. [CrossRef] [PubMed]

109. Conklin, H.M.; Ashford, J.M.; Clark, K.N.; Martin-Elbahesh, K.; Hardy, K.K.; Merchant, T.E.; Ogg, R.J.; Jeha, S.; Huang, L.; Zhang, H. Long-term efficacy of computerized cognitive training among survivors of childhood cancer: A single-blind randomized controlled trial. *J. Pediatr. Psychol.* **2016**, *42*, 220–231. [CrossRef] [PubMed]

110. Chacko, A.; Bedard, A.; Marks, D.; Feirsen, N.; Uderman, J.; Chimiklis, A.; Rajwan, E.; Cornwell, M.; Anderson, L.; Zwilling, A. A randomized clinical trial of Cogmed working memory training in school-age children with ADHD: A replication in a diverse sample using a control condition. *J. Child Psychol. Psychiatry* **2014**, *55*, 247–255. [CrossRef] [PubMed]

111. Patel, S.K.; Ross, P.; Cuevas, M.; Turk, A.; Kim, H.; Lo, T.T.; Wong, L.F.; Bhatia, S. Parent-directed intervention for children with cancer-related neurobehavioral late effects: A randomized pilot study. *J. Pediatr. Psychol.* **2014**, *39*, 1013–1027. [CrossRef] [PubMed]

112. Holland, A.A.; Hughes, C.W.; Harder, L.; Silver, C.; Bowers, D.C.; Stavinoha, P.L. Effect of motivation on academic fluency performance in survivors of pediatric medulloblastoma. *Child Neuropsychol.* **2016**, *22*, 570–586. [CrossRef] [PubMed]

113. Riggs, L.; Piscione, J.; Laughlin, S.; Cunningham, T.; Timmons, B.W.; Courneya, K.S.; Bartels, U.; Skocic, J.; de Medeiros, C.; Liu, F. Exercise training for neural recovery in a restricted sample of pediatric brain tumor survivors: A controlled clinical trial with crossover of training versus no training. *Neuro-Oncology* **2016**, *19*, 440–450. [CrossRef] [PubMed]

114. Conklin, H.M.; Reddick, W.E.; Ashford, J.; Ogg, S.; Howard, S.C.; Morris, E.B.; Brown, R.; Bonner, M.; Christensen, R.; Wu, S. Long-term efficacy of methylphenidate in enhancing attention regulation, social skills, and academic abilities of childhood cancer survivors. *J. Clin. Oncol.* **2010**, *28*, 4465. [CrossRef] [PubMed]

115. Castellino, S.M.; Tooze, J.A.; Flowers, L.; Hill, D.F.; McMullen, K.P.; Shaw, E.G.; Parsons, S.K. Toxicity and efficacy of the acetylcholinesterase (AChe) inhibitor donepezil in childhood brain tumor survivors: A pilot study. *Pediatr. Blood Cancer* **2012**, *59*, 540–547. [CrossRef] [PubMed]

116. Boele, F.W.; Douw, L.; de Groot, M.; van Thuijl, H.F.; Cleijne, W.; Heimans, J.J.; Taphoorn, M.J.; Reijneveld, J.C.; Klein, M. The effect of modafinil on fatigue, cognitive functioning, and mood in primary brain tumor patients: A multicenter randomized controlled trial. *Neuro-Oncology* **2013**, *15*, 1420–1428. [CrossRef] [PubMed]

117. Brown, P.D.; Pugh, S.; Laack, N.N.; Wefel, J.S.; Khuntia, D.; Meyers, C.; Choucair, A.; Fox, S.; Suh, J.H.; Roberge, D. Memantine for the prevention of cognitive dysfunction in patients receiving whole-brain radiotherapy: A randomized, double-blind, placebo-controlled trial. *Neuro-Oncology* **2013**, *15*, 1429–1437. [CrossRef] [PubMed]

118. Poggi, G.; Liscio, M.; Pastore, V.; Adduci, A.; Galbiati, S.; Spreafico, F.; Gandola, L.; Massimino, M. Psychological intervention in young brain tumor survivors: The efficacy of the cognitive behavioural approach. *Disabil. Rehabil.* **2009**, *31*, 1066–1073. [CrossRef] [PubMed]

119. Barrera, M.; Atenafu, E.G.; Sung, L.; Bartels, U.; Schulte, F.; Chung, J.; Cataudella, D.; Hancock, K.; Janzen, L.; Saleh, A. A randomized control intervention trial to improve social skills and quality of life in pediatric brain tumor survivors. *Psycho-Oncology* **2018**, *27*, 91–98. [CrossRef] [PubMed]

120. Jackson, A.C.; Enderby, K.; O'Toole, M.; Thomas, S.A.; Ashley, D.; Rosenfeld, J.V.; Simos, E.; Tokatlian, N.; Gedye, R. The role of social support in families coping with childhood brain tumor. *J. Psychosoc. Oncol.* **2009**, *27*, 1–24. [CrossRef] [PubMed]

121. Macartney, G.; Stacey, D.; Harrison, M.B.; VanDenKerkhof, E. Symptoms, coping, and quality of life in pediatric brain tumor survivors: A qualitative study. *Oncol. Nurs. Forum* **2014**, *41*, 390–398. [CrossRef] [PubMed]

122. Ris, M.D.; Grosch, M.; Fletcher, J.M.; Metah, P.; Kahalley, L.S. Measurement of neurodevelopmental changes in children treated with radiation for brain tumors: What is a true 'baseline?'. *Clin. Neuropsychol.* **2017**, *31*, 307–328. [CrossRef] [PubMed]

123. Kesler, S.R.; Gugel, M.; Huston-Warren, E.; Watson, C. Atypical structural connectome organization and cognitive impairment in young survivors of acute lymphoblastic leukemia. *Brain Connect.* **2016**, *6*, 273–282. [CrossRef] [PubMed]

© 2018 by the authors. Licensee MDPI, Basel, Switzerland. This article is an open access article distributed under the terms and conditions of the Creative Commons Attribution (CC BY) license (http://creativecommons.org/licenses/by/4.0/).

MDPI

St. Alban-Anlage 66

4052 Basel

Switzerland

Tel. +41 61 683 77 34

Fax +41 61 302 89 18

www.mdpi.com

Bioengineering Editorial Office

E-mail: bioengineering@mdpi.com

www.mdpi.com/journal/bioengineering

www.ingramcontent.com/pod-product-compliance
Lightning Source LLC
Chambersburg PA
CBHW051910210326
41597CB00033B/6097